20 084 184 56

AF606193

NANOTECHNOLOGY SCIENCE AND TECHNOLOGY

REMOTE SENSING

TECHNIQUES, APPLICATIONS AND TECHNOLOGIES

NANOTECHNOLOGY SCIENCE AND TECHNOLOGY

Additional books in this series can be found on Nova's website under the Series tab.

Additional e-books in this series can be found on Nova's website under the e-book tab.

NANOTECHNOLOGY SCIENCE AND TECHNOLOGY

REMOTE SENSING

TECHNIQUES, APPLICATIONS AND TECHNOLOGIES

ENNER ALCÂNTARA
EDITOR

New York

For permission to use material from this book please contact us:
Telephone 631-231-7269; Fax 631-231-8175
Web Site: http://www.novapublishers.com

NOTICE TO THE READER

Additional color graphics may be available in the e-book version of this book.

Library of Congress Cataloging-in-Publication Data

Remote sensing : techniques, applications and technologies / editor, Enner Alcbntara.
pages cm.
Includes index.
ISBN 978-1-62417-140-6 (hardcover)
1. Remote sensing. I. Alcbntara, Enner.
G70.4.R4688 2013
621.36'78--dc23

2012038768

Published by Nova Science Publishers, Inc. † New York

Contents

PREFACE

As many authors has defined Remote Sensing, the classical one can be, the science and the art of obtaining information about an object, area or phenomenon through the analysis of data acquired by a device that is not in a direct contact with the target.

This book tries to show some techniques and applications of remote sensing in sorted themes, such as: coastal basins, biomass estimation, water resources management, particle dispersion in hydroelectric reservoirs and urban growth. Also, this book shows an interesting way to study the attitude control of satellite. It is separated into six chapters.

Chapter 1 - 'Land-based remote sensing of coastal basins: use of an HF radar to investigate surface dynamics and transport processes in the Gulf of Naples'. This chapter investigates the surface dynamics of a coastal area using a radar. The radar data are integrated with transport model to estimate the residence times of the particles and to identify possible areas of retention and aggregation.

Chapter 2 - 'Remote sensing applications of estimating biomass for energy crops-development of ground-based sensing systems'. This chapter focused in techniques to understand crop status from different sensors including multispectral camera, spectrometer, laser range finder and light quantum sensors.

Chapter 3 - 'Remote sensing as a support tool for water resources management in Sub-Saharan Africa: a review'. This chapter shows some techniques for hydrological modeling, flood forecasting, water conservation and efficiency in irrigation.

Chapter 4 - 'Estimating particle dispersion and reservoir transport processes from satellite tracked drifters data: methodology and applications'.

This chapter shows techniques to study the horizontal transport and dispersion in large hydroelectric reservoirs through the use of satellite tracked drifters.

Chapter 5 - 'Remote sensing and attitude control using intelligent image-based sensor'. The main objective of this chapter is to use an imaging camera as an intelligent device for extracting relative attitude information needed for the attitude control.

Finally, the chapter 6 -'Assessing urban growth and soil sealing with spectral mixture analysis'. In this chapter the authors shows a remote sensing techniques to study the urban growth over time, combining unsupervised, supervised classifications and the spectral mixture analysis.

In: Remote Sensing
Editor: Enner Alcântara

ISBN: 978-1-62417-140-6

Chapter 1

LAND-BASED REMOTE SENSING OF COASTAL BASINS: USE OF AN HF RADAR TO INVESTIGATE SURFACE DYNAMICS AND TRANSPORT PROCESSES IN THE GULF OF NAPLES

Daniela Cianelli[1,2], Marco Uttieri[1], Raffaele Guida[1], Milena Menna[3], Berardino Buonocore[1], Pierpaolo Falco[1], Giovanni Zambardino[1] and Enrico Zambianchi[1]

[1]Università degli Studi di Napoli "Parthenope", Dipartimento di Scienze per l'Ambiente, Napoli, Italy

[2]Istituto Superiore per la Protezione e la Ricerca Ambientale,Roma, Italy

[3]IstitutoNazionale di Oceanografia e Geofisica Sperimentale, Sgonico (TS), Italy

ABSTRACT

Here, the results of an investigation that was performed on the surface dynamics of a coastal area using data provided by a land-based remote sensing system, a HF coastal radar, are presented. The surface circulation in the Gulf of Naples (southern Tyrrhenian Sea, western

Mediterranean), as determined using measurements collected by a HF radar, displayed recurrent dynamics directly linked to prevailing forcing conditions. In this contribution we examine the circulation-dependent surface transport of passive particles released in a coastal sub-basin (Bay of Naples), as characterized by severe anthropic pressure. HF radar data of surface circulation were integrated using a transport model, and simulations of coastal-offshore transport were performed in order to estimate the residence times of the particles and to identify possible areas of retention and aggregation. Four test cases, representative of the most frequent seasonal observed circulation structures, were investigated. Depending on the forcing factor driving surface circulation, the results emphasized the existence of unique mechanisms.

1. INTRODUCTION

The subtitle of Ian Robinson's most recent book "Discovering the Oceans from Space" (2010) reads "The Unique Applications of Satellite Oceanography". The main aspect of uniqueness is the capability of remote sensing techniques to provide synoptic (i.e. simultaneous, over a broad area) measurements of physical and biological quantities within the upper layers of the ocean. If taking synoptic snapshots can only be obtained from a distance (*in situ* measurements are, by definition, not synoptic unless massive arrays of sensors are employed, which in the ocean has proved not to be feasible in logistic or economic terms), the possibilities are not only limited to satellite remote sensing. As early envisioned by Sherman (1969 and 1978), broader synoptic coverage can be fulfilled by aircraft-mounted, as well as by ship- or land-based oceanographic sensors.

Over the last two decades interest has arisen in HF coastal radars, a type of land-based remote sensing system that provides high spatial and temporal resolution measurements of surface currents in coastal basins. The availability of synoptic current measurements not only allows detailed studies of ocean circulation, but may also provide advective fields for models that predict the transport of pollutants or trajectories in search and rescue operations.

The history of land-based HF radar applications for ocean current mapping spans more than four decades. Since HF radar applications were designed for military purposes (mainly for detecting hard targets: ships, aircraft and missiles, and even Kelvin wakes from ships), the origin of the use of such oceanographic equipment for this purpose is almost accidental. The HF radar systems of the past were very bulky and extremely expensive to operate, and were characterized by very strong background noise due to backscattering from the sea surface (for strategic reasons they were typically installed in coastal regions). A breakthrough occurred when the perspective was inverted, and people began to consider backscattering from the sea surface as the signal rather than as the noise (Pederson and Barrick, 2004). Technology developments have allowed antenna downsizing (the original antennas installed by Don Barrick on San Clemente Island in 1970 were 500 m long!) and systems are now extremely compact, allowing for very easy installation in diverse coastal settings. As a result, the number of systems installed along the ocean's coasts has strongly increased over the last 10-20 years (Harlan et al., 2010). Here, we demonstrate and exploit the capabilities of a HF coastal radar installed in the Gulf of Naples, a marginal basin of the southern Tyrrhenian Sea (western Mediterranean Sea).

Historical reports on the circulation of the Gulf of Naples (henceforth GoN) highlight the occurrence of highly dynamical structures over the basin (Moretti et al., 1976-1977 and 1983; De Maio et al., 1985; Cianelli et al., 2012; and for a recent assessment of summer circulation in the GoN see Uttieri et al., 2011). As for other marginal and coastal basins, even in GoN, wind stress represents the most important local forcing, directly and indirectly – e.g., through interactions with local orography and bottom topography - affecting surface current fields, as evident from both empirical and modelling results (Moretti et al., 1976-1977; De Maio et al., 1985; Gravili et al., 2001; Grieco et al., 2005; Menna et al., 2007; Cianelli et al, 2012). Wind forcing displays a typical seasonality (Menna et al., 2007; Uttieri et al., 2011) that is associated with the large-scale atmospheric circulation acting over the region. When such forcing is weak, as often in summer, a locally induced breeze regime can act as the primary driver of sea surface circulation.

The circulation of the GoN can also be remotely driven by the larger-scale circulation of the southern Tyrrhenian Sea (Gravili et al., 2001), which it is in communication with through the "Bocca Grande" and "Bocca Piccola" apertures (Figure 1). The Tyrrhenian circulation has a marked barotropic component (Pierini and Simioli, 1998) and is basically organized in three main gyres (two cyclonic and a central anticyclonic) with typical seasonal patterns (Artale et al., 1994). Recent observations from surface drifters and satellite altimetry (Rinaldi et al., 2010) have shown that circulation in the southern Tyrrhenian Sea is extremely complex, and only in a very broad sense can be considered as "overall cyclonic" (e.g., Millot, 1999). However, it is modulated by a series of mesoscale/subbasin structures, of both a transient and semi-permanent nature. The importance of these structures overcomes the mean flow. In fact, during winter an intense northward flow may develop along the eastern Tyrrhenian coast, as supported by the intrusion of Modified Atlantic Water (Krivosheya and Ovchinnikov, 1973), resulting in a coastal current that represents the main remote forcing of the circulation in the interior of the GoN.

In addition to its unique hydrographic and oceanographic features, for socio-economic reasons the GoN has a leading role among marginal basins of the southern Tyrrhenian Sea. Severe anthropic activity impinges the area and ranges from urban settlements to tourism and maritime traffic that can have drastic consequences on ecosystem dynamics and water quality. The necessity of preserving and sustainably exploiting environmental resources requires an accurate understanding and the monitoring of circulation and transport processes acting within the basin, by forestalling or mitigating potentially hazardous events. In recent years, increasing awareness regarding the importance of operational oceanography systems (e.g., Flemming et al., 2002; Dahlin et al., 2003) has promoted the development of new techniques and instruments to investigate real-time, multi-scale circulation patterns in a target area as well as their effects on transport and pollutant diffusion, with a special emphasis on coastal zone management.

The purpose of this work was to study surface circulation patterns within the GoN as driven by typical local wind conditions and by southern Tyrrhenian guidance, and to assess their effects on the transport processes of passively buoyant particles. The synoptic, basin-scale circulation of the GoN was reconstructed using a HF radar system that has been operating since October 2004. From the available dataset we selected four periods, each lasting one week, with typical forcing conditions (section 2). For each period we investigated the associated surface current velocity and the vorticity fields (section 3). Thereafter, to simulate the fate of passive conservative particles released in a coastal sub-area, the Bay of Naples (section 4), we applied a transport model. The results presented here indicate that each set of conditions determined unique circulation structures and transport processes, as the result of the superimposition of multiple co-occurring factors. Such an integrated approach, recently utilized for the study of a very specific pollution event in the area (Uttieri et al., 2011), allowed an evaluation of surface water renewal times and investigations regarding exchange mechanisms between sub-basins.

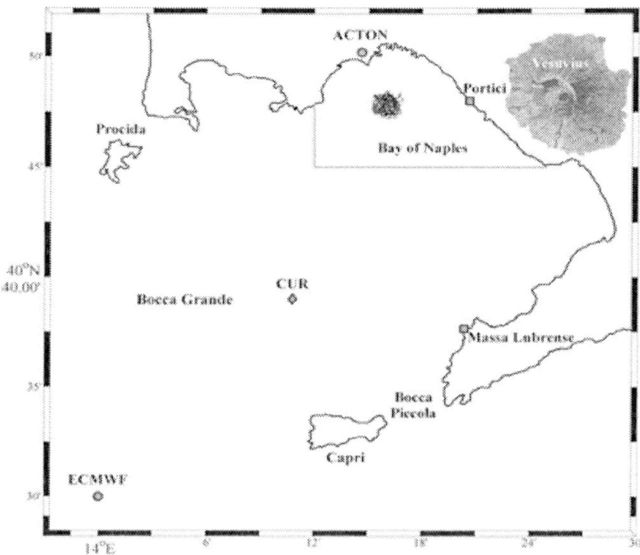

Figure 1. A map of the Gulf of Naples (southern Tyrrhenian Sea), with the location of the two radar antennas whose data were used in this work (Portici and Massa Lubrense), the DiSAm Acton weather station, the ECMWF grid point and the point where currents were evaluated in the breeze regime case (CUR, see Figure 4).

2. Local and Remote Forcings of the Surface Circulation

The analysis of local wind forcing was performed using the meteorological data set recorded by a weather station located in the urban area of Naples (Acton station: 40° 30'.19 N, 14° 15'.21 E). The station belongs to the monitoring network managed by the Department of Environmental Sciences at Parthenope University (DiSAm). To compare small-scale with basin-scale wind features, data relative to the closest gridpoint (40° 30'.00 N, 14° 00'.00 E) of the operational weather forecast model of the European Centre for Medium-Range Weather Forecasts (ECMWF) were also analyzed (Figure 1). The Acton station and ECMWF data were in good directional agreement; only in cases of southwesterly wind ECMWF recordings was a shift in direction determined, most likely due to the position of the Acton station which, owing to orographic barriers, is sheltered from westerly winds (Menna et al., 2007).

Interannual investigations over the 2002-2006 period (Menna et al., 2007) indicated the occurrence of two prevailing wind directions, one from the NE-NNE and one from the SW-SSW, in direct association with large-scale meteorological dynamics. When these large-scale forcings are weak, as often recorded during summer, a breeze regime arises and winds blow from the sea to land during the day and from land to sea during the night. The breeze regime can also be observed during other periods of the year, typically in correspondence to passages of high pressure centers within the GoN.

Using these results, we selected three characteristic periods from the available data set, each lasting one week, during which the wind field maintained marked directional stability. The chosen time span allowed us to analyze the formation and the development of the surface circulation structures associated with different forcing conditions, as follows:

- *The LND case study:* winds from the NE (Figure 2a). The selected period corresponded to November 1-8 2007, with wind intensities in the range 2-6 m s^{-1} and occasional gusts ≥ 8 m s^{-1}.
- *The SEA case study:* winds from the SW (Figure 3a). The selected period for this condition was January 17-24, 2007. The stick diagram in Figure 3a shows wind velocities in the range 2-6 m s^{-1}, with occasional gusts ≥ 12 m s^{-1}.
- *The BRZ case study:* the breeze regime (Figure 4a). The selected period for this situation was August 1-8, 2007. The stick diagram shows alternating wind directions approximately every 12 hours with lower intensities than in the previous cases (0-6 m s^{-1}).

In addition to the three case studies discussed above, we also investigated surface circulation as directly guided by a current in the neighbouring southern Tyrrhenian Sea. Preliminary investigations (De Maio et al., 1983; Cianelli et al., 2012) indicated that the direct effect of the southern Tyrrhenian is manifested through the development of an intense jet aligned along the Capri/Ischia direction, enhancing the intrusion of Tyrrhenian waters into the GoN. The result is supported by the analysis of trajectories of drifting buoys launched in the southern Tyrrhenian and entering the basin (Rinaldi et al., 2010). Therefore, in this work we also present the following fourth case:

- *The TYR case study:* southern Tyrrhenian forcing (Figure 5a). The period selected was December 10-17, 2006. At this time of the year the southern Tyrrhenian circulation displays an intense northward flux (Artale et al., 1994; Pierini and Simioli, 1998). The recordings for the selected period show intense south-easterly winds, with occasional changes in direction.

3. Surface Dynamics As Detected by a Land-Based HF Coastal Radar

The acronym RADAR (RAdio Detection And Ranging) defines an apparatus that sends an electromagnetic signal, typically within the range of microwaves, toward a target and receives its backscattered echo. From the time lag between transmission and reception it is possible to calculate the target distance.

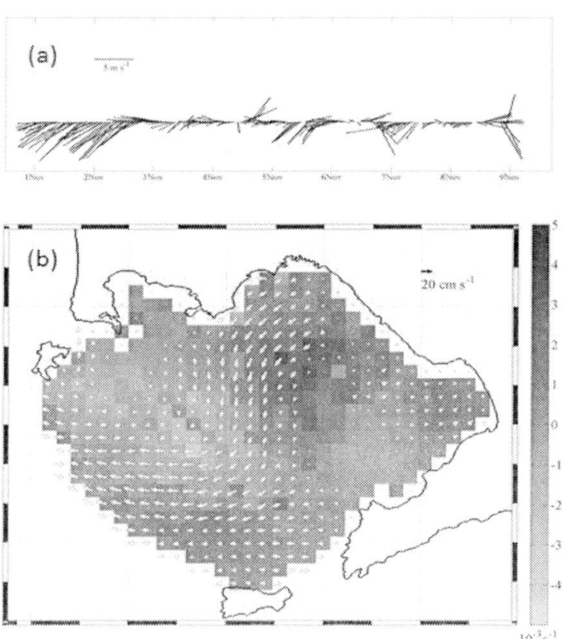

Figure 2. The LND case study: **(a)** a stick diagram for Acton wind data; **(b)** a surface current and vorticity map (Figure 2b refers to the circulation observed on Nov 5th, 2007, at 16:00 GMT).

For a coastal radar the target is represented by the crests and troughs of the gravity waves propagating at the surface of a coastal basin. The applicability to such a system is a consequence of Bragg scattering, originally conceived to explain the diffraction of X rays by a crystal lattice, which can be applied to a sea surface perturbed by the presence of gravity waves. The sea surface may act as a diffraction grating for the electromagnetic signal at a suitable wavelength. The backscattered radiation will be in a condition of constructive interference when the wavelength of the sea surface roughness is approximately half the wavelength of the incoming signal.

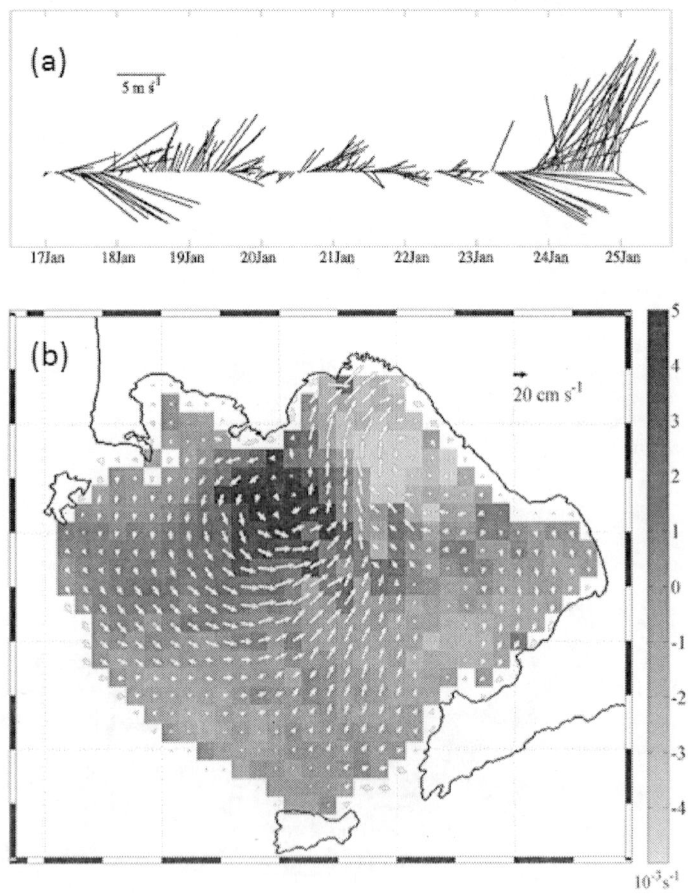

Figure 3. The SEA case study: **(a)** a stick diagram for Acton wind data; **(b)** a surface current and vorticity map (Figure 3b refers to the circulation observed on Jan 17th, 2007, at 18:00 GMT).

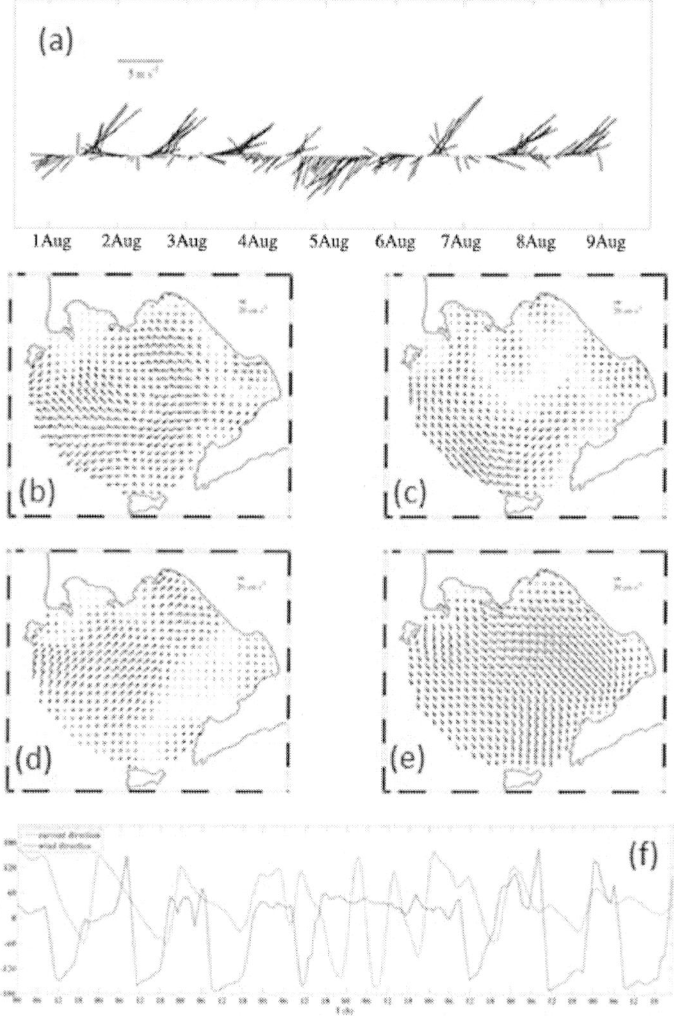

Figure 4. The BRZ case study: **(a)** a stick diagram for Acton wind data; **(b-e)** surface currents measured on Aug 1st, 2006, at 00:00, 06:00, 12:00, 18:00 GMT respectively; **(f)** the wind and current (measured at point CUR as shown in Figure 1) direction.

The spectrum of gravity waves that develop in coastal basins have relatively universal characteristics. Therefore, the operational frequency of coastal radar is typically within a range that optimizes the probability of backscattering by surface waves. The choice of operating frequencies for individual systems can be determined by considering its sought-after range and resolution.

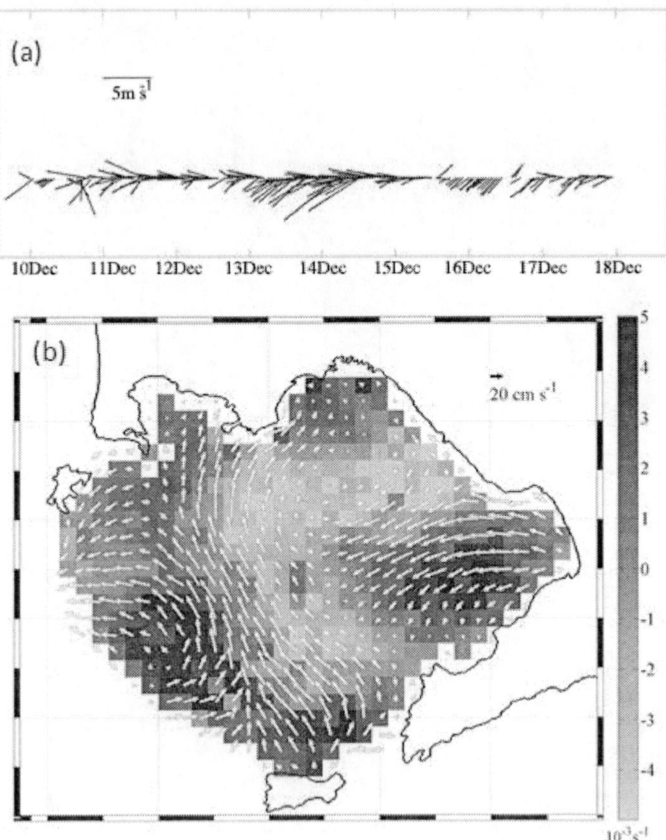

Figure 5. The TYR case study: **(a)** a stick diagram for Acton wind data; **(b)** a surface current and vorticity map (Figure 5b refers to the circulation observed on Dec 17th, 2006, at 09:00 GMT).

If the surface gravity waves responsible for backscattering are stationary waves, the signal received from each station will have the theoretical characteristics described above. However, in reality waves are almost never stationary, and propagate toward or away from the antenna location; yielding a frequency variation for the received signal, due to the Doppler effect, that causes a frequency increase for approaching wave trains and a decrease for retracting ones.

If waves are superimposed onto a surface current field, the frequency of the backscattered signal is characterized by an additional Doppler shift that is induced by the radial (approaching or retracting from the receiving station) component of the current (Barrick et al., 1977) that can be utilized to evaluate the current. When at least two transceiving stations are available, it is possible to combine their radial information on surface currents and to obtain a vector current field (Barrick and Lipa, 1986). Exhaustive reviews on HF radar principles and functioning can be found in Paduan and Graber (1997) and Teague et al. (1997).

The synoptic surface circulation in the GoN was analyzed using data provided by a HF coastal radar (a SeaSonde system, manufactured by CODAR O.S. of Mountain View, California) that has been operated in the area since 2004 by the DiSAm on behalf of the AMRA consortium (where AMRA stands for the Analysis and Monitoring of Environmental Risks).

The data presented in this work refers to the network configuration active until 2008, composed of two monostatic, direction-finding systems with three-element crossed loop/monopole antenna used to transmit and receive signals. To determine the bearing of an incoming signal, the directional properties of the antennas are exploited and a variant of the multiple signal classification algorithm is applied (MUSIC, see Schmidt, 1986). Antennas were located in Portici and Massa Lubrense, while the central site for data processing and merging was hosted at DiSAm headquarters. The nominal operating frequency of the systems was 25 Mhz, and provided hourly data with a 1.250 km spatial resolution (the technical specifications of the utilized system are provided in Table 1; Figure 1 shows the antenna locations; a third antenna at Castellammare has recently been added, bringing the resolution to 1 km).

Table 1. The technical specifications of the HF radar system installed in the Gulf of Naples

RADIAL MAP ACQUISITION	
Frequency	25 Mhz
Sweep Rate	2.01 Hz
Samples per Sweep	2048
Band Width	150.147 Khz
Range Step	0.999 km
Range	~ 30.0 km
Wavelength	12 m
RADIAL MAP COMBINE OPTIONS	
Grid Spacing	1.250 km
Averaging Radius	2.0 km
Distance Angular Limit	25° GDOP limit
Current Velocity Limit	80 cm s^{-1}

Coastal radar coverage is often impacted by spatial and/or temporal gaps (see, e.g., Paduan and Cook, 1997). In particular, due to the presence of a Saracen tower in the immediate vicinity of the Massa Lubrense antenna, our system had a blind region in the southeastern portion of the GoN. For this reason, to interpolate the data in space so as to fill the no-data region with geometrical/dynamical methods, an Open-boundary Mode Analysis (OMA) procedure (Lekien et al., 2004) was adopted.

The OMA is an evolution of the Normal Mode Analysis (NMA) method introduced in the study of coastal circulation by Lipphardt et al. (2000), and represents its improvement in terms of the representation of the flow at open boundaries. Both the OMA and NMA are spectral objective mapping techniques that use two subsets of functions (vorticity modes and divergence modes) to represent a velocity field. The functions exactly enforce a no normal flow condition at the domain's solid boundaries and use a separate inhomogeneous open boundary solution to account for specified normal flow through the domain's open boundaries. The functions produce maps that are precisely consistent with a three-dimensional incompressible velocity field (the method can be extended to use 3D mapping functions to directly map 3D velocity data sets that are dense enough to properly constrain the mapping).

CODAR current data have also been used to reconstruct the vorticity field associated with surface circulation by means of the relative vorticity (ζ) evaluated as the vertical component of the curl of the velocity field (Pedlosky, 1979), as follows:

$$\zeta = (\vec{\nabla} \times \vec{v})_z = \frac{\partial v}{\partial x} - \frac{\partial u}{\partial y}.$$

In the following, we briefly describe the circulation observed in the GoN for the four case studies.

The LND case study: In the presence of winds from the NE, surface currents (Figure 2b) are roughly oriented along the E-W direction and describe a jet (mean velocity ~ 30 cm s^{-1}) in the central-southern region of the basin that is flanked by cyclonic and anticyclonic vorticity structures. The occurrence of the coastal jet as typically associated with northeasterly winds has been observed in the past using experimental data (Moretti et al., 1976-1977; Menna et al., 2007), as well as in model results (Gravili et al., 2001; Grieco et al., 2005).

The SEA case study: For SW wind conditions a coastward flux develops from the Island of Capri to the coast and to the Bay of Naples, with the formation of a persistent cyclonic gyre in the offshore sector of the GoN and the presence of other recirculation structures (both cyclonic and anticyclonic) at the basin and sub-basin scale (Figure 3b). Toward the end of the investigated period, the gyre progressively reduced its intensity until it disappeared, while a net coastward current spread over the entire basin. The results are in full agreement with previous observations (Menna et al., 2007) and confirm the numerical simulations of Gravili et al. (2001) and Grieco et al. (2005).

The BRZ case study: During the breeze regime, surface currents respond to wind stress with a periodic oscillation; the surface current pattern makes a 360° clockwise rotation within 24 hrs (Figures 4b-f). The wind intensity follows a daily breeze trend: the highest intensity was reached each day around h 12:00-13:00, while the lowest intensity was reached during night hours around midnight.

The TYR case study: For this case surface circulation in the GoN is guided by Tyrrhenian input that penetrates from the "Bocca Piccola" and partially from the "Bocca Grande", as also verified by drifting buoy data (Rinaldi et al., 2010). When the Tyrrhenian meridional flow converges in the GoN, the surface circulation in its interior is characterized by a jet current (with a mean velocity of ~ 12 cm s^{-1}) oriented from the SE to the NW along the Bocca Piccola - Procida Channel axis (Figure 5b). The jet current axis separates the GoN circulation into two parts, as follows: east to the surface current describes an anticyclonic circulation, while on the west it is cyclonic. The patterns are in agreement with the previous *in situ* observations of De Maio et al. (1983).

The above description suggests obvious qualitative agreement between wind regimes and current patterns (with the exception of the *TYR* case study), and was further investigated and quantitatively assessed by performing a cross-correlation between winds and currents.

In order to have more than one wind data point in the area, we resorted to the Cross-Calibrated Multi-Platform (CCMP) surface wind velocities made available by the NASA Physical Oceanography Distributed Active Archive Center. The data are derived from ocean surface wind data from SSM/I, TMI, AMSR-E, SeaWinds on QuikSCAT, and SeaWinds on ADEOS-II, combined with conventional observations; and from a starting estimate of the wind field using a variational analysis method. Wind fields were available from 1987 to 2010 every 6 hours with a 25 km spatial resolution (Ardizzone et al., 2009).

For each period, wind data were interpolated in space and time on the HF radar data grid. In order to estimate Ekman currents, a complex linear regression model was applied, as follows:

$$\boldsymbol{U} = \beta e^{i\vartheta}\boldsymbol{W} + error$$

where **U** and **W** are the radar current and CCMP are the wind velocities, expressed as complex numbers (**W** = w_1+iw_2), β is a real constant, and θ is an angle. β and θ represent the speed factor and the rotation angle of the Ekman currents with respect to the wind direction. Similar models have been applied to drifter data (e.g., by Niiler et al., 2003; Poulain et al., 2009; and Poulain et al., 2012) and also to radar data in the Kuroshio area (Tokeshi et al., 2007). The results of this analysis are summarized in Table 2. The Ekman currents estimated from the coastal radar were 1.3 to 2.0% of the wind speed, rotated to the right of the wind direction with an angle varying between 21° and 26°.

The complex correlation coefficient, evaluated according to Kundu (1990), provided information on the correlation between currents and winds. In Table 2, the correlation is expressed in terms of the determination coefficient (R^2), which is the square of the complex correlation coefficient (Emery and Thomson, 2001). In the *LND* case study the Ekman current was responsible for 3% of the variance of surface currents measured by coastal radar. In the *SEA* case study the percentage rose to 11%, and in the *BRZ* case study to 21%, whereas in the *TYR* case study it was reduced to 0.5%. The result confirms the qualitative discussion and indicates how the highest correlation occurs with a breeze, whereas the correlation is basically null when

Table 2. Results of the analysis of the cross correlation between Ekman and total surface currents for the four case studies. See section 3 for an explanation. N represents the total number of data points over which the analysis was performed

Case study	Wind direction	Bexp(θ)	R^2 (%)	N
LND	NE	0.013exp(-21°i)	3	68734
SEA	SW	0.014exp(-25°i)	11	78995
BRZ	Breeze	0.020 exp(-23°i)	21	63907
TYR	NE	0.014exp(-26°i)	0.5	76355

a strong Tyrrhenian coastal current develops. For the case of steady and strong winds (for the case studies *LND* and *SEA*) the correlation was half as high as that of the breeze case. For those cases, the locally forced circulation is apparently modulated by more complex dynamics. The difference between the *LND* and *SEA* case studies most likely rests in the fact that winds from the NE were distorted by the presence of Vesuvius, which has an influence on a portion, but possibly not the entire, basin.

4. Tracer Transport in the Gulf of Naples

The GoN represents an area of great interest to tourists and for commercial activities, but is extremely vulnerable (from a marine environment standpoint) due to the presence of several industrial sites along the coast, highly polluted river estuaries, and intense maritime traffic. For this reason, an understanding of the transport dynamics developing in the basin is important for implementing response strategies and procedures in cases of emergencies, whether they be represented by a pollution event, an oil spill, or a shipwreck.

Surface transport processes of passive tracers have been studied in the GoN by applying an oil spill model, the GNOME trajectory model (General N.O.A.A. Operational Modeling Environment) developed by the National Oceanic and Atmospheric Administration (NOAA) of the U.S.A.. GNOME is a mixed Lagrangian-Eulerian transport model that provides particle trajectory estimates by processing oceanographic as well as atmospheric data for a fixed geographic region (Beegle-Krause, 2001). GNOME offers three modes that differ from one another in the degree of control on the input parameters. In our study we used GNOME as a classic transport model in the diagnostic mode, and set all of the model data fields and parameters (Beegle-Krause, 2001). We considered a Lagrangian passive and conservative tracer moved by GNOME for an Eulerian current and wind field. Surface current fields were supplied to the model by the HF radar system operating within the GoN. HF radar current estimates present intrinsic uncertainties as compared to *in situ* measurements (e.g., Chapman et al., 1997; Ullman and Codiga, 2004). To account for this effect, we set a ± 15% error estimate in GNOME in both the along- and cross-current directions.

As an additional forcing mechanism contributing to transport processes ("mover", in GNOME jargon), we used hourly-averaged 10 m wind data collected by the DiSAm using an automatic weather station located on the urban littoral of Naples (Acton, see Figure 1). As discussed in section 2, since wind is a major driver of circulation in the GoN, its effect on circulation is already included in the surface current field detected by the radar system. However, in the simulations the direct effect of wind on the movement of a surface buoyant body, as could occur for an oil spill or for the case of a search and rescue simulation, was included. In GNOME this is accomplished by including the *windage* parameter, by including the Stokes drift and the momentum directly imparted by wind onto the particle (Engie and Klinger, 2007).

In addition to advection by surface currents and wind drift, a stochastic component to particle motion, in the form of a random walk (see, e.g., Csanady, 1980) parameterized by eddy diffusivity, was also included.

The relative effects of advection, windage, and diffusion were also studied in a parametric way by keeping the current constant and by allowing the wind and diffusivities to span the ranges reported in Table 3. In particular, for the wind field we used the actual value measured by the weather station, one tenth of the value, and also considered the case of a total absence of wind. We let the diffusivity range from zero to 10^6 cm^2 s^{-1} using 10^4 and 10^5 cm^2 s^{-1}. Worth noting, here, is that typical diffusivity values for dynamics with a spatial scale on the order of 1-10 km can be estimated to occur between 10^4 and 10^5 cm^2 s^{-1} (Okubo, 1971).

For the four case studies introduced in the previous sections, transport processes were numerically investigated using GNOME. Simulations of the trajectories of 10,000 particles were performed for each one-week period.

The study of transport processes in the GoN have largely focused on the properties of its most important sub-area immediately off the urban littoral (the Bay of Naples (see Figure 1)), in particular in regards to the effectiveness of its water renewal mechanisms. Therefore, particles have been deployed in the Bay and particle exchange between the Bay and the interior of the Gulf have been assessed in order to describe the effects of advection, diffusion, and wind in terms of two integral descriptors of transport within a semi-enclosed basin, as defined by Buffoni et al. (1997) for the normalized tracer quantity in the Bay $Q(t)$ and the tracer residence time T.

Table 3. The forcings used for the parametric study. In the wind intensity column "v" represents the speed measured at the Acton station

WIND INTENSITY	DIFFUSIVITY(cm^2 s^{-1})
0, v/10, v	0, 10^4, 10^5, 10^6

The normalized number of particles that, at a generic time (t) following deployment, were still located inside the release area (Bay of Naples) was estimated as follows:

$$Q(t) = \frac{N(t)}{N(0)}$$

where $N(t)$ is the number of particles present at time t and $N(0)$ is the pool of the originally launched tracers.

When applicable, the residence time, the time spent by a particle in the release area before exiting, was computed as follows:

$$T = \lim_{t \to \infty} T^*(t)$$

with:

$$T^*(t) = t\,Q(t) + \sum_{i=1}^{N_e(t)} \frac{t_{ei}}{N(0)}$$

where $N_e(T) = N(0) - N(t)$ is the number of particles that had already abandoned the deployment area at t, and t_{ei} represents the time taken by the ith particle to first exit. Preliminary analysis indicated that when a particle exited the original domain it was not retained subsequent to the time period investigated, making the definition for a buffer zone (as outlined, e.g., by Falco et al., 2000) unnecessary.

Figures 6, 7, 8, and 9 display the behavior of Q(t) for all of the performed experiments, as well as for the final particle distribution in the AD case (see below) for the most realistic level of horizontal diffusion (10^4 cm^2 s^{-1}).

Given the number of different experiments performed (12 forcing combinations for four different case studies, leading to 48 simulations), in the following we focus our discussion on three forcing combinations for each case study. Thereafter, we briefly comment on the results contained in Table 4, which provides the values attained for the residence time (T) for all of the experiments.

In the description below, we define A as the case for which transport is only due to advection by surface currents (i.e. the wind and diffusion were set to zero). AD is defined as the case for which both advection and diffusion (with diffusivity set to 10^4 cm^2 s^{-1}) were present. ADW is the case for which wind was measured by the weather station in addition to advection and diffusion.

The LND case study (Figure 6)*:* The LND case study was characterized by a very effective renewal mechanism of Bay of Naples waters, and in all simulations the final concentration of the tracer particles in the bay did not exceed 4%. For the A condition all of the particles abandoned the release sector and remained afloat, with a T ~44 h. The cloud was split into two parts after ~3 days: one was advected by the coastal jet that rapidly brought particles toward the offshore sector of the GoN, while the other was transported to coastal sub-areas (Gulf of Castellammare) where local circulation entrapped the tracer.

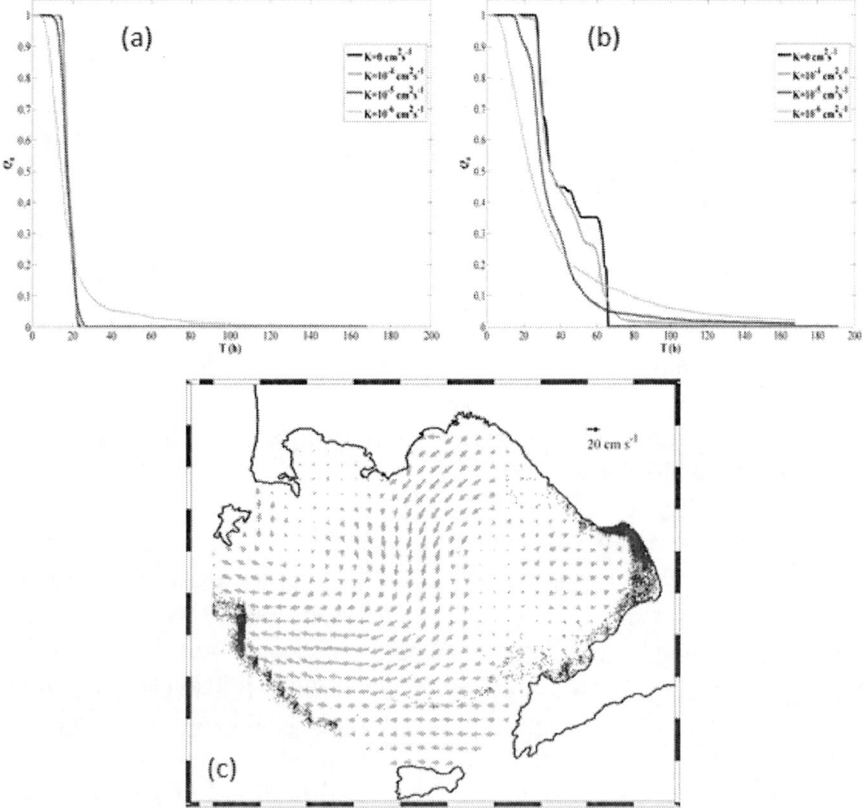

Figure 6. Results of the transport simulation relative to the LND case study: **(a)** Q(t) for null wind; **(b)** Q(t) for actual wind as measured by the Acton weather station; and **(c)** the final particle distribution for the case of null wind, with a diffusivity of K=10^4 cm^2 s^{-1}**.**

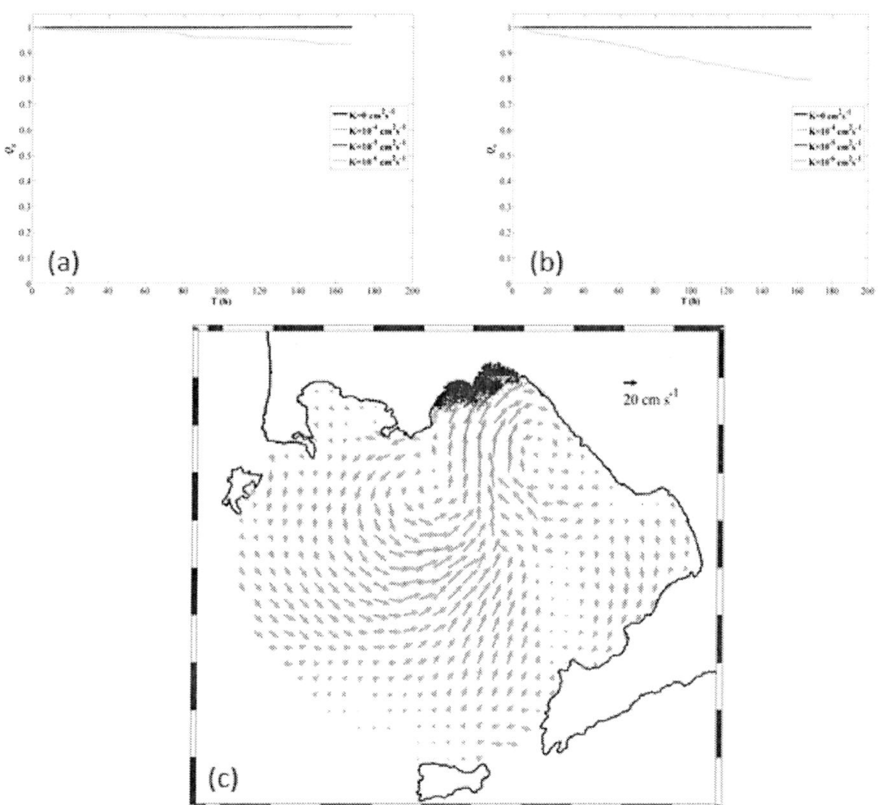

Figure 7. Results of the transport simulation relative to the SEA case study: **(a)** Q(t) for null wind; **(b)** Q(t) for actual wind as measured by the Acton weather station; and **(c)** the final particle distribution for the case of null wind, with a diffusivity of $K=10^4$ cm^2 s^{-1}.

In the AD case, *T* was also ~44 h. Diffusion spread the cloud which, in this case, separated into two branches after ~3 days. One of the clouds entered the coastal jet and approached the "Bocca Grande". Approximately 1% of the particles abandoned the GoN at the end of the simulation. In the second branch, the vast majority of the particles remained floating in the Gulf of Castellamare and in nearby areas, with a small percentage (12%) becoming stranded.

For the ADW condition all of the particles left the Bay of Naples, with a *T* of ~19 h. The wind rapidly pushed the tracer which, in ~4 days, reached the outer portion of the domain.

The SEA case study (Figure 7)*:* SW winds induced the strong retention of tracer particles within the Bay of Naples, and in all of the mover configurations *T* definitely exceeded the duration of the simulation. As compared to A, the effect of diffusion (AD) was manifest through the spreading of the cloud, with 24% of particles being stranded by the end of the simulation. When windage was accounted for (ADW), diffusion became the factor that massively (81%) pushed the tracer along the coast.

The BRZ case study (Figure 8)*:* For the A condition all of the particles remained inside the release area. Within 24 hours, most of them were advected toward the coast by the sea breeze regime, where they became trapped in a small area without reaching open waters. In the AD case, particles became distributed inside the release area and were alternatively moved toward the coast or to open waters according to the breeze regime. By the 5th day of the simulation, some of the particles were stranded inside the launch sector (11%), while others were advected outside of the launch sector. Particles were transported by a meandering structure that retained them inside the GoN.

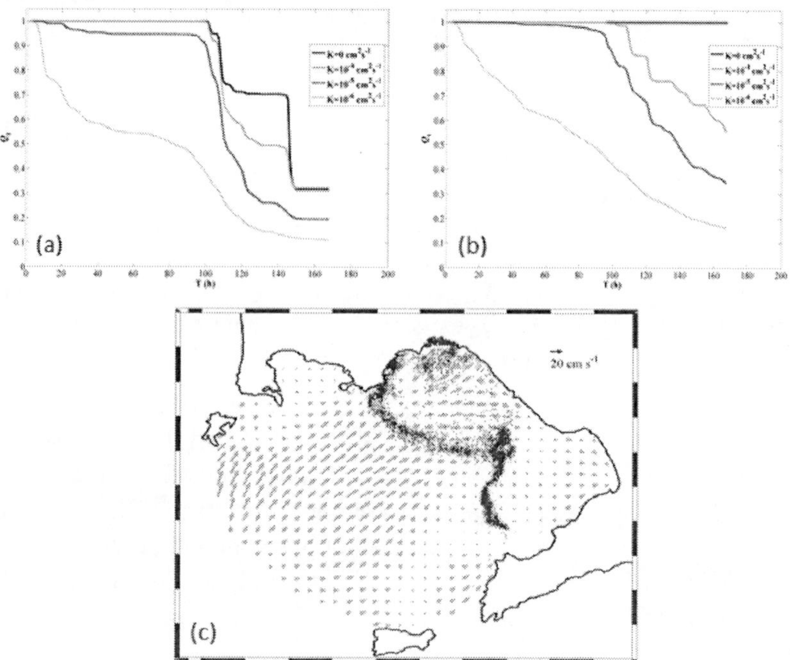

Figure 8. Results of the transport simulation relative to the BRZ case study: **(a)** Q(t) for null wind; **(b)** Q(t) for actual wind as measured by the Acton weather station; and **(c)** the final particle distribution for the case of null wind, with a diffusivity of $K=10^4$ cm^2 s^{-1}.

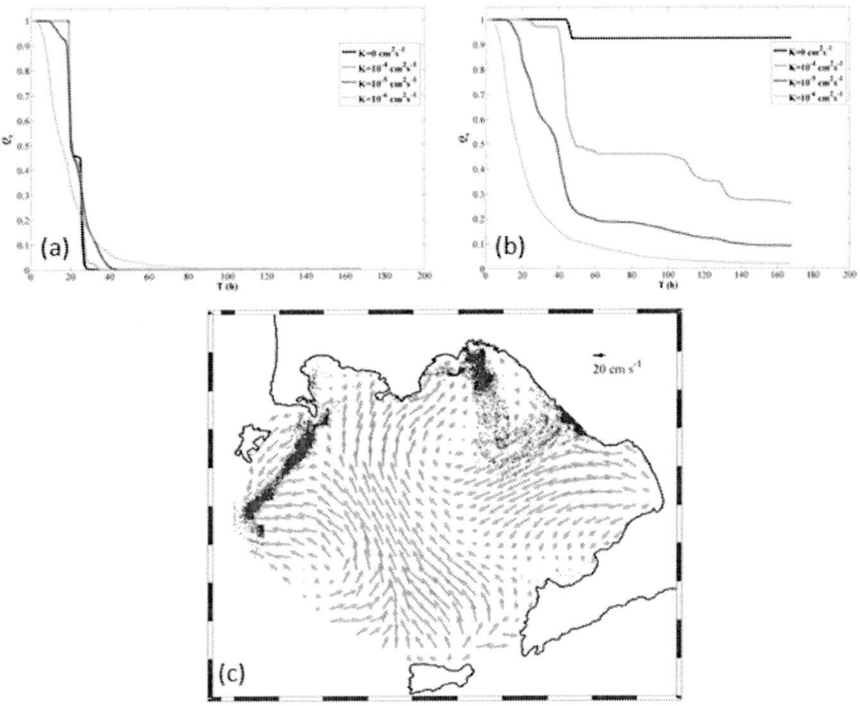

Figure 9. Results of the transport simulation relative to the TYR case study: **(a)** Q(t) for null wind; **(b)** Q(t) for actual wind as measured by the Acton weather station; and **(c)** the final particle distribution for the case of null wind, with a diffusivity of $K=10^4$ $cm^2\ s^{-1}$.

Similar results were reported for the ADW condition. The distinguishing feature was represented by a larger amount of beached particles inside the release area (26%).

The TYR case study (Figure 9): For case A the tracer was first advected toward the boundary of the release sector within ~24 hours. Then it was divided into two small clusters. The first was transported by a coastward current until stopping in an area not sampled by the radar system. The second was advected toward the offshore sector. After roughly 5 days, a portion entered a coastal sub-area (Bay of Pozzuoli), whereas the remaining one crossed the Tyrrhenian jet and entered the cyclonic circulation on the left of the jet, approaching the "Bocca Grande".

The overall dynamics were much more complex for the AD condition. When the cluster split after ~24 h, a first subset migrated toward the Bay of Pozzuoli. The second subset underwent more intricate dynamics. Part of it was first moved toward the coast. However, a meandering flow then transported most of the particles offshore. Therefore, another splitting was observed. Many of the tracers entered the Bay of Naples again and approached the littoral zone. Others, instead, crossed the Tyrrhenian jet and entered an offshore cyclonic gyre, being moved out of the GoN (22%). A similar scenario also occurred when windage was taken into account.

Table 4. The residence times, *T*, estimated in simulations for different forcing choices

Case Study LND			
Wind forcing diffusivity	*Calm*	*V/10*	*V*
0	44 hrs	32 hrs	19 hrs
10^4 cm^2s^{-1}	44 hrs	33 hrs	19 hrs
10^5 cm^2s^{-1}	37 hrs	31 hrs	18 hrs
10^6 cm^2s^{-1}	36 hrs	33 hrs	19 hrs
Case Study SEA			
Wind forcing diffusivity	*Calm*	*V/10*	*V*
0	› 168 hrs	› 168 hrs	› 168 hrs
10^4 cm^2s^{-1}	› 168 hrs	› 168 hrs	› 168 hrs
10^5 cm^2s^{-1}	› 168 hrs	› 168 hrs	› 168 hrs
10^6 cm^2s^{-1}	› 168 hrs	› 168 hrs	› 168 hrs
Case Study BRZ			
Wind forcing diffusivity	*Calm*	*V/10*	*V*
0	› 168 hrs	163 hrs	146 hrs
10^4 cm^2s^{-1}	156 hrs	157 hrs	138 hrs
10^5 cm^2s^{-1}	138 hrs	138 hrs	120 hrs
10^6 cm^2s^{-1}	87 hrs	87 hrs	74 hrs
Case Study TYR			
Wind forcing diffusivity	*Calm*	*V/10*	*V*
0	› 168 hrs	45 hrs	23 hrs
10^4 cm^2s^{-1}	92 hrs	52 hrs	23 hrs
10^5 cm^2s^{-1}	52 hrs	55 hrs	23 hrs
10^6 cm^2s^{-1}	27 hrs	27 hrs	19 hrs

From the results for all of the simulations summarized in Table 4, we can see that in the *LND* case study, in which currents are strong and the dynamics relatively steady, the influence of diffusion on the tracer residence time in the Bay of Naples was small. On the contrary, the effect of windage strongly impacted estimates by enhancing water renewal by more than 2x in terms of *T*.

Also, for the *SEA* case study that was characterized by strong currents and steady dynamics, diffusion did not make a significant difference. In this case this is true also for wind, obvious since its enhancing effect cannot be quantified in such a retentive (for the Bay of Naples) configuration.

For the *BRZ* case study that was dominated by a breeze regime, windage made only weak differences in the renewal mechanism of the Bay, and was strongly enhanced by the presence of diffusion that diminished *T* by a factor of two within the considered diffusivity range.

Diffusivity was also very important for the *TYR* case study, especially for weak or null wind. For the latter case, diffusivity decreased *T* by at least one order of magnitude. On the other hand, when windage was fully taken into account, given the fact that the dominating winds during that week blew from the NE, the results indicate a very strong enhancement of renewal, whether or not diffusion was present.

Summary and Concluding Remarks

Here, the results of an investigation on surface currents, as well as the associated transport processes within the Gulf of Naples, as detected by a land-based remote sensing system, a HF coastal radar, are presented. Coastal radar provides a unique opportunity for obtaining a synoptic view of surface currents in relatively wide basins (with measurement ranges reaching 200 km for the case of long-range instrument configurations), with a high spatial and temporal resolution. For investigations of coastal dynamics, their characteristics make them almost irreplaceable for studies of pollutant transport or for search and rescue operations. For this reason, these systems are becoming more and more widespread along the world's coastlines (see, e.g., Harlan et al., 2010).

Synoptic measurements of surface currents allowed us to first identify typical circulation patterns within the GoN, strictly subordinate to local wind-stress or remote forcing. The relative importance of local and remote forcing was assessed by means of a cross-correlation analysis.

Wind in the area has been shown to mainly blow from the NE and SW quadrants, with the exception of summer which is dominated by a breeze regime. The three forcing configurations typically originate from surface circulation patterns, which have been investigated in dynamical terms, and with regard to the effectiveness of coastal water renewal, with a focus on the Bay of Naples - immediately off the urban littoral.

In the presence of NE winds, basin scale circulation develops a jet located in the central region of the GoN. The jet structure, directed offshore towards the Bocca Grande, typically induces cyclonic and anticyclonic circulations on a sub-basin scale. Also, in the southern region of the basin, surface currents are oriented along the wind direction. The resulting circulation promotes offshore-oriented transport from the coastal sector and prevents the import of particles from the offshore sector toward the Bay of Naples.

For SW winds, the GoN surface circulation is, on the whole, characterized by onshore dynamics. In the coastal areas of the basin, cyclonic and anticyclonic structures are formed, and support water recirculation processes, preventing the renewal of water masses. In the northern portion of the basin, surface circulation is more structured and presents an intense coastward oriented current. The area maintains a clear tendency to cyclonic rotation. Such surface dynamics enhance the retention of particles within the coastal area, as well as with an onshore particle flux from the interior of the Gulf.

In the breeze regime, the surface current direction closely follows the evolution of the wind direction. The entire pattern of surface circulation is seen to rotate clockwise, completing a 360° rotation in 24 hrs. The current pattern supports the gradual particle emptying of the coastal zone, while only a limited fraction of the particles released in the center of the Gulf reach the Bay of Naples.

In addition to locally forced regimes, the presence of a strong current at the outer boundary of the GoN may become a remote forcing, inducing a circulation in the interior that may be independent of the wind. In such a situation, Tyrrhenian water masses flow into the GoN and canalize through the Bocca Piccola forming a jet current oriented from SE to NW. The jet structure separates the GoN dynamics in two distinct sectors - east of the Tyrrhenian jet the surface current assumes an anticyclonic orientation, whereas on the west a cyclonic circulation develops. Both structures and the jet are minimally correlated with the wind direction, indicating that such dynamics are not driven by local wind forcing but are instead influenced by the Southern Tyrrhenian Winter circulation. The Tyrrhenian-driven circulation is favourable in regards to the renewal of Bay of Naples surface waters.

The relative importance of the different factors affecting surface transport in the area was assessed by examining the residence time of tracer particles using different simulations for an assortment of movers - strong and steady winds also led to strong and steady current patterns yielding residence times that were not sensitive to the effect of diffusion. On the contrary, transport processes in the case of a breeze regime were strongly impacted by diffusivity, and the case of remote forcing was seen to strongly depend on additional windage effects.

Our work reaffirms the unique possibilities provided by current measurements obtained using remote sensing instruments such as land-based HF coastal radars, that allow a synoptic view that is crucially important for studying both local dynamics and transport processes in coastal areas. These systems may play a leading role in operational oceanography networks – as recently demonstrated on the occasion of the environmental catastrophic event of the Deepwater Horizon oil spill in the Gulf of Mexico during the spring of 2010, making them invaluable to decision makers and stakeholders in devising specific policies for the management of coastal zones.

ACKNOWLEDGMENTS

The Dipartimento di Scienze per l'Ambiente of the "Parthenope" University operates the HF radar system on behalf of the AMRA consortium (formerly CRdC AMRA), a Regional Competence Center for the Analysis and Monitoring of Environmental Risks. Our radar remote sites were hosted by the ENEA centre of Portici, the "Villa Angelina Village of High Education and Professional Training" and "La Villanella" resort in Massa Lubrense, and the Fincantieri shipyard in Castellammare di Stabia, whose hospitality is gratefully acknowledged. Our work was partly funded by the MED TOSCA project, co-financed by the European Regional Development Fund; by the PROMETEO project, funded by the Campania Region; by the National Programme of Scientific and Technological Research RItMare. Preparatory work was carried out in the framework of the EU Interreg III B Archimed CORI (the Prevention and Management of Sea Originated Risk to the Coastal Zone) project, and the Italian MIUR funded VECTOR project (subtasks 4.1.5 and 4.1.6). Partial support from the Procura della Repubblica (Public Prosecutor's office) of Torre Annunziata is also acknowledged.

We thank Jeffrey D. Paduan and Michael S. Cook for providing the OMA routines. Jeffrey D. Paduan's advice on several issues related to the coastal radar setup and data processing is also gratefully acknowledged. Discussions with Pierre-Marie Poulain helped improve estimates of Ekman currents. Gennaro Bianco is thanked for providing Acton wind data, and Alessandro Mercatini and Pasquale Mozzillo are thanked for supportive collaboration.

The GNOME package is freely available from NOAA's Office of Response and Restoration.

REFERENCES

Ardizzone, Atlas, Hoffman, Jusem, Leidner and Moroni.(2009). New multiplatform ocean surface wind product available.EOS Transactions American Geophysical Union.90, 231.

Artale, Astraldi, Buffoni and Gasparini.(1994). Seasonal variability of gyre-scale circulation in the Northern Tyrrhenian Sea. *Journal of Geophysical Research.* 99 (C7), 14127-14137.

Barrick and Lipa. (1986). An evaluation of least-squares and closed-form dual-angle methods for CODAR surface-current applications. *IEEE Journal of Oceanic Engineering.* OE-11, 322-326.

Barrick, Evans and Weber.(1977). Ocean surface currents mapped by radar. *Science.* 198, 138-144.

Beegle-Krause. (2001) General NOAA Oil Modeling Environment (GNOME): a new spill trajectory model. In *IOSC 2001 Proceedings*; Mira Digital Publishing, Inc.: Tampa, FL, pp 865-871.

Buffoni, Falco, Griffa and Zambianchi. (1997). Dispersion processes and residence times in a semi-enclosed basin with recirculating gyres: an application to the Tyrrhenian Sea. *Journal of Geophysical Research.* 102 (C8), 699-713.

Chapman, Shay, Graber, Edson, Karachintsev, L. and Ross. (1997). On the accuracy of HF radar surface current measurements: intercomparisons with ship-based sensors. *Journal of Geophysical Research.* 102 (C8), 18737-18748.

Cianelli, Uttieri, Buonocore, Falco, Zambardino and Zambianchi. (2012) Dynamics of a very special Mediterranean coastal area: the Gulf of Naples. In *Mediterranean Ecosystems: Dynamics, Management and Conservation;* Williams (Ed.); Nova Science Publishers, Inc.: New York, pp. 129-150.

Csanady. (1980). *Turbulent Diffusion in the Environment;* D. Reidel Publishing Company: Dordecht, pp 260.

ɾsson. (2003). *Building the European* *ography.* Elsevier Oceanography Series, ›p 714.

zie and Vultaggio.(1983). Dinamica delle ˌdiacenze. *Risultati ottenuti dal* 1977 al 1981. › Navale - Napoli. LI, 1-58.

pezie and Vultaggio. (1985). Outline of marine Naples and some considerations on pollutant *nto.* 8C, 955-969.

(2001). *Data Analysis Methods in Physical* vier: Amsterdam, pp 654.

007). Modeling passive dispersal through a large system to evaluate marine reserve network connections. *Estuaries and Coasts*. 30, 201-213.

Falco, Griffa, Poulain and Zambianchi.(2000). Transport properties in the Adriatic Sea as deduced from drifter data. *Journal of Physical Oceanography*. 30, 2055-2071.

Flemming, Vallerga, Pinardi, Behrens, Manzella, Prandle and Stel.(2002). *Operational Oceanography - Implementation at the European and Regional Scales*. Elsevier Oceanography Series, Vol. 66; Elsevier: Amsterdam, pp 572.

Gravili, Napolitano and Pierini.(2001). Barotropic aspects of the dynamics of the Gulf of Naples (Tyrrhenian Sea). *Continental Shelf Research*. 21, 455-471.

Grieco, Tremblay and Zambianchi. (2005). A hybrid approach to transport processes in the Gulf of Naples: an application to phytoplankton and zooplankton population dynamics. *Continental Shelf Research*. 25, 711-728.

Harlan, Terrill, Hazard, Keen, Barrick, Whelan, Howden and Kohut. (2010). The integrated ocean observing system high-frequency radar network: status and local, regional, and national applications. *Marine Technology Society Journal*. 44, 122-132.

Krivosheya and Ovchinnikov.(1973). Peculiarities in the geostrophic circulation of the waters of the Tyrrhenian Sea. *Oceanology*. 13, 822-827.

Kundu. (1990). *Fluid Mechanics*; Academic Press: San Diego, pp 730.

Lekien, Coulliette, Bank and Marsden. (2004). Open boundary modal analysis: interpolation, extrapolation, and filtering. *Journal of Geophysical Research*. 109, C12004.

Lipphardt, Kirwan, Grosch, Lewis and Paduan.(2000). Blending HF radar and model velocities in Monterey Bay through normal mode analysis. *Journal of Geophysical Research.* 105 (C2), 3425-3450.

Menna, Mercatini, Uttieri, Buonocore and Zambianchi. (2007). Wintertime transport processes in the Gulf of Naples investigated by HF radar measurements of surface currents. Il *Nuovo Cimento*. 30 C, 605-622.

Millot.(1999). Circulation in the Western Mediterranean Sea. *Journal of Marine Systems*. 20, 423-442.

Moretti, Spezie and Vultaggio. (1983). Sub-inertial waves observed in the Gulf of Naples. *Rapports de la Commission Internationale pour l'Exploration Scientifique de la Mer Méditerranée*. 28, 151-154.

Moretti, Sansone, Spezie, Vultaggio and De Maio. (1976-1977). Alcuni aspetti del movimento delle acque del Golfo di Napoli. Annali, Istituto Universitario Navale - Napoli. XLV-XLVI, 207-217.

Niiler, Maximenko, Panteleev, Yamagata and Olson. (2003). Near-surface dynamical structure of the Kuroshio Extension. *Journal of Geophysical Research.* 108, 3193.

Okubo.(1971). Oceanic diffusion diagrams. *Deep Sea Research*. 18, 789-802.

Paduan and Cook. (1997). Mapping surface currents in Monterey Bay with CODAR-type HF radar. *Oceanography*. 10, 49-52.

Paduan and Graber. (1997). Introduction to High-Frequency radar: reality and myth. *Oceanography*. 10, 36-39.

Pederson and Barrick.(2004). HF surface-wave radar-revisiting a solution for EEZ ship surveillance. *EEZ International*. 35-37.

Pedlosky. (1979). *Geophysical Fluid Dynamics*; Springer-Verlag: New York, pp 624.

Pierini and Simioli.(1998). A wind-driven circulation model of the Tyrrhenian Sea area. *Journal of Marine Systems*. 18, 161-178.

Poulain, Menna and Mauri. (2012). Surface geostrophic circulation of the Mediterranean Sea derived from drifter and satellite altimeter data. *Journal of Physical Oceanography*. 42, 973-990.

Poulain, Gerin, Mauri and Pennel.(2009). Wind effects on drogued and undrogued drifters in the Eastern Mediterranean. *Journal of Atmospheric and Oceanic Technology*. 26, 1144-1156.

Rinaldi, Buongiorno Nardelli, Zambianchi, Santoleri and Poulain. (2010). Lagrangian and Eulerian observations of the surface circulation in the Tyrrhenian Sea. *Journal of Geophysical Research*. 115 C04024.

Robinson. (2010). *Discovering the Ocean from Space - The Unique Applications of Satellite Oceanography;* Springer: London, pp 638.

Schmidt. (1986). Multiple emitter location and signal parameter estimation. *IEEE Transactions on Antennas and Propagation*. 34, 276-280.

Sherman. (1969). Synoptic Oceanography: Remote Sensing Studies of the Ocean; *DTIC Document: Fort Belvoir*, pp 20.

Sherman. (1978). An Overview of Remote Sensing Oceanogrpahy in the United States. *Proceedings of International Symposium on Remote Sensing of Environment (12th),* Held at Ann Arbor, MI. on Apr 20-26, 1978; National Technical Information Service: Alexandria, pp 14.

Teague, Vesecky and Fernandez.(1997). HF radar instruments, past to present. *Oceanography*. 10, 40-44.

Tokeshi, Ichikawa, Fujii, Sato and Kojima. (2007). Estimating the geostrophic velocity obtained by HF radar observations in the upstream area of the Kuroshio. *Journal of Oceanography*. 63, 711-720.

Ullman and Codiga. (2004). Seasonal variation of a coastal jet in the Long Island Sound outflow region based on HF radar and Doppler current observations. *Journal of Geophysical Research*. 109, C07S06.

Uttieri, Cianelli, Buongiorno Nardelli, Buonocore, Falco, Colella and Zambianchi. (2011). Multiplatform observation of the surface circulation in the Gulf of Naples (Southern Tyrrhenian Sea). *Ocean Dynamics*. 61, 779-796.

In: Remote Sensing
Editor: Enner Alcântara

ISBN: 978-1-62417-140-6

Chapter 2

Remote Sensing Applications of Estimating Biomass for Energy Crops: Development of Ground-Based Sensing Systems

Tofael Ahamed[1,*], Lei Tian[2], Noguchi Ryozo[1] and Tomohiro Takigawa[1]

[1]Graduate School of Life and Environmental Sciences, University of Tsukuba, Tsukuba, Japan

[2]Department of Agricultural and Biological Engineering, University of Illinois at Urbana-Champaign, IL, US

Abstract

Biomass from energy crops is a very versatile resource that can be used to produce heat, electricity, and transport fuels. The energy crops have high potentials to use in all these applications today, and its future demand is projected in regional and national levels to grow substantially. To maximize the biomass feedstock production, remote sensing techniques have been identified for the non-destructive measurements and estimation of biomass. The physical status of crop growth and biomass accumulation can be projected over the growing seasons. Thus, the

* Corresponding author: Tofael Ahamed, 1-1-1 Tennodai Tsukuba, 305-8572, Ibaraki, Japan. Tel: +81-29-853-8835, Fax: +81-29-853-4922 Email: tofael.ahamed.gp@u.tsukuba.ac.jp.

objective of this chapter is to focus the techniques of understating crop status from different sensors including multispectral camera, spectrometer, laser range finder and light quantum sensors. These non-destructive measurements include estimation of biomass correlating with suitable vegetative indices, height of the plants, and biomass accumulation using Photosynthetically Active Radiation (PAR). The field experimental data has also presented from field spectrometry and light quantum sensing to show the potentials of measuring suitable vegetative indices, and methods of estimating the amount of dry matter biomass from PAR. The image information captured from the top of a 38 m tower using a multispectral camera for energy crops helped to understand the crop growth status over the growing season. The Vegetation Index (VI), and intercepted PAR of energy crops have been measured at the Energy Farm, University of Illinois at Urbana-Champaign to develop a model for predicting dry matter biomass. While conducting the experiments in the bioenergy crop's field, the using of spectrometry was difficult due to height of the energy crops. Therefore, a high clearance Data Acquisition Vehicle (DAV) was proposed for ground sensing of energy crops. In an another field experiment, at the Agricultural and Forest Research Center, University of Tsukuba, a laser range finder was used to determine the plants growing height from biomass canopy to estimate the yield of biomass as an non-destructive remote measurement from an autonomous agricultural tractor. This laser based approach was a step to develop an on-the-go remote sensing system under the tree canopy where illumination is a concern. The remote sensing applications in the estimation of biomass, and energy content using multiple sensors could be installed with a microcontroller to develop an on-the-go sensing platform and asses the real-time crop growing status.

1. INTRODUCTION

Remote sensing is the key technology for site-specific management of crop production. Researchers have developed precision agricultural technologies and processes that have enhanced agricultural production in traditional crops like corn and soybeans. Now the trends have been changed due to practices of bioenergy crops to maximize biomass feedstock production throughout the world.

Biomass feedstock production is a critical subsystem within the overall bio-based energy production and utilization system. It provides necessary materials input to the conversion process of biomass into fuel, power, and value-added materials. This subsystem includes the operations of agronomic production of energy crops and physical handling/delivery of biomass, as well as other enabling logistics. (Ting et al, 2008). The agronomic production depends on yield variability over the growing season and utilizes the optimum harvesting window.

The measurement of yield variability of biomass feedstock is needed for developing and evaluating site-specific crop management practices in their perennial growth. In different growth stages, the field spectroscopy has the fundamental importance for assessing spectral response of plant canopies and photosynthetically active radiation for biomass conversion and multispectral imagery are the key points to understand crop response in remote sensing applications. The field spectroscopy involves the study of interrelationships between the spectral characteristics of objects and their biophysical attributes in the field environment.

It acts firstly, as a bridge between the laboratory measurements of spectral reflectance and the field situation and is useful in calibration of airborne and satellite sensors. Secondly, it is useful in predicting the optimum spectral bands viewing configuration and time to perform a particular remote sensing task. Thirdly, it provides a tool for the development, refinement, testing of models relating biophysical attributes to remotely sensed data (Milton, 1987). The light interception indicates the conversion of solar energy into biomass which is cheap but not very efficient process: about 2% of the crop captured incoming energy within the photosynthetically active spectrum of solar radiation is converted into biomass (Vargas et al. 2002).

The multispectral imagery refers images capture data at specific wavelengths across the electromagnetic spectrum. The wavelengths may be separated by filters or by the use of instruments that are sensitive to particular wavelengths, including light from frequencies beyond the visible light range, such as infrared. Multi-spectral imagery can allow extraction of information from spectral response that the human eye fails to capture with its receptors for red, green and blue. The relationships between crop reflectance in the visible and near infrared wave length are closely correlated with the amount of photo synthetically active tissue in the crop (Jørgensen et al. 2003).

The most widely accepted method for describing vegetative growth using reflectance spectra is band ratios or vegetation indices. Vegetation indices are spectrally-based values generated through the mathematical manipulation of reflectance measurements from two or more spectral wavelengths (Thorp, 2004). The vegetation index is use to quantify the concentrations of green leaf vegetation (Lyon et al., 1998). Linear combination from two or more wavebands may be more sensitive and robust to assess the crop status than a signal band (Bausch and Duke, 1996). Generally, vegetation indices can be divided into broadband indices and narrowband indices according to the bandwidth of image data.

The broadband indices are calculated based on broadband reflectance data and the narrowband indices are calculated using narrow spectral bands acquired by a spectrometer or a hyperspectral image sensor (Yao, 2003). There are more than 20 broadband vegetation indices that have been designed to represent different crop information from remote sensing images (Jensen, 2000). The Normalized Difference Vegetation Index (NDVI) is the most commonly used vegetation index and Gitelson et al. (1996) proposed the Green Normalized Difference Vegetation Index (GNDVI) which substituted the red band in the NDVI with the green band. The GNDVI proves to be more useful for assessing canopy variation in green crop biomass.

The vegetative indices are the indicators from reflectance measurements and could be used to correlate with dry matter estimation for perennial grasses. Therefore this study attempted to find correlation between field spectrometric data with tower based multispectral imagery to track vegetative response over the growing season and estimate biomass from light interception. The correlation between biomass and suitable vegetative indices was figured out in the growing seasons. Again, in precision agriculture, crop height and biomass are important parameters to assess the crop yield parameters. One of the potential biomass crops is Miscanthus, which has a high yield, perennial growth and good resistance against disease, cold and drought.

To ensure proper growth of Miscanthus, it is essential to know the plant stress, fertilization timing, physical parameters, and soil environment. It is also important to monitor and observe these parameters over the growing season. Miscanthus grows higher and denser as the growing season progresses. Two-month-old Miscanthus grows faster, and the height of these plants is approximately 50 cm. The 3-year Miscanthus grows up to 3 meters high. However, data acquisition is difficult due to lack of high clearance vehicle operating as on-the-go sensing system for Miscanthus and other biomass feedstock.

Currently, aerial hyperspectral and multispectral images are available for agricultural remote sensing to find nitrogen stress and mapping (Yao and Tian 2004; Bajwa and Tian 2001; Yang et al 2002; Xiang and Tian 2011). On the contrary, the illumination had an effect in the remote measurements of optical sensors. In these cases, commercially available low cost laser range finders have been investigated in both horticultural and agricultural research, for example, canopy volume and structure measurement in citrus (Tumbo et al.

2002). Besides, optical sensor, laser based integrated systems would help in monitoring biomass and height of the crops together from the relative measurement of reflective cloud scan points. From tractor-based platform, the combined approach for growth monitoring using LRF has got significant importance to carry out in the outdoor operations due to low cost and suitability while illumination is a concern (Ahamed et al. 2009, Ahamed et al. 2011).

Therefore, the independent monitoring systems and data acquisition platforms are required for estimating biomass over the growing season. Hence, the objectives of this research are to develop optimized instrumentations and data processing systems for ground-based remote sensing applications to monitor perennial growth of biomass feedstock.

2. Materials and Methods

2.1. Development of Stand-Alone Sensing

2.1.1. Image Acquisition and Instrumentations

The stand-alone camera sensor system was developed with a 4 bands MS4100, a multispectral charged couple device (CCD) camera (Geospatial), a pan/tilt device (PT570P medium duty) and receiver (LRD41C21/22 Legacy® and a lens controller (Figure 1-a). The multispectral camera was a digital progressive scan camera with a high resolution of 1920x1080 pixels.

In contrast to a normal CCD camera, the camera was available in two spectral configurations: RGB for high quality color imaging and color-infrared for multispectral applications. The camera had 3 CCD channels and channels had center wavelengths of 650nm, 800nm and 500nm, respectively and bandwidth of approximately 100nm for each. A serial interface provided external control of gain and exposure time for each independent channel via a standard RS 232 port. The gain settings controlled the amount of the output signal amplification for each individual channel in the camera. The gain of the camera ranged from 0dB to 36dB corresponding to 95 to 1023 in 16-bit digital number representation and 928 steps in total. Exposure time was the amount of time that each channel in the camera accumulates charge before the electronic shutter was closed and resulting value was read out.

The exposure time of the camera varies from 0.1ms to 108 ms that corresponds to 16-bit digital number from 1 to 1080, 1079 steps in total. The maximum frame rate of the camera was 10 frames per second. The camera was able to output 8-bit and 10-bit digital image for each channel. The 8-bit mode and a digital frame grabber IMAQ PCI 1428 (National Instruments, Austin, TX) was used in the image acquisition process. The PCI 1428 was a highly flexible IMAQ board for PCI chassis that supports a diverse range of Camera Link-compatible camera. The PCI 1428 had been installed into an industrial small rugged computer with PCI expansion slot (SC241S) that could operate at extreme outdoor conditions.

A serial port of the computer was connected to the external control port of the camera via a nine-pin serial cable. The pan/tilt device was rotated in horizontal and vertical directions to get the images according to the plot distributions. The Pelco D protocol was used to communicate with the pan tilt device and receiver using RS232 serial communication. The pan tilt rotates 0º-355º horizontally and 0º-90º degree vertically. The presets according to the field distribution were established using the caller identifications and automatic rotations of the pan/tilt device had been developed. The lens motorization was developed externally and used two motors to control zoom and focus.

The calibration was performed for zoom and focus of the lens using two potentiometers. The camera sensor system was developed to capture images from 38 meter height tower for Miscanthus, switch grass, prairie grass and corn (Figure 1-b, c).

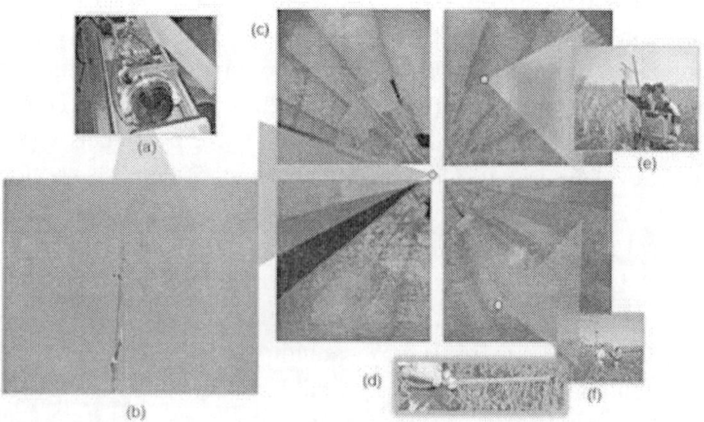

Figure 1. Tower remote sensing system and ground reference points data collections using spectrometer (a) Field layout with mosaic image (b, c) Spectral sensing using spectrometer (d) Light quantum sensing, (e) Stand-alone tower remote sensing (f) multispectral camera.

The average plot size was 9 acres and stand-alone tower was erected at the middle of the four fields. The perennial crops were at the first year of their growth and needed to replant rhizomes for uniform density of canopy. The perennial energy crop like Miscanthus requires 3 to 4 years to establish in a uniform pattern.

2.2. Ground-Reference Field Sampling

During photosynthesis, plants used energy in the region of the electromagnetic spectrum from 400-700 nm. The radiation in this range, referred to as Photosynthetically Active Radiation (PAR), could be measured in energy units or as Photosynthetic Photon Flux Density (PPFD), which had units of quanta (photons) per unit time per unit surface area. The scaled units most commonly used were micromoles of quanta per second per square meter (μmol $s^{-1}m^{-2}$). A LI-COR 191SA line quantum sensor was used in a horizontal position (Figure 1-d).

Incoming photo synthetically active radiation (PAR) was measured with the sensor held upright above the canopy at an approximate height of 1.5 m above the soil surface. Reflected PAR was measured with the sensor held inverted above the canopy at an approximate height of 0.5 m above the canopy surface. Transmitted PAR was measured with the sensor held upright in the canopy at the soil surface. The sensor had a near hemispherical field-of-view along its 1 m length. The LI-COR 191SA Quantum sensor absolute calibration was within +/- 5% (usually with +/- 3% under most sky conditions.

It was cosine corrected up to an 80° solar zenith angle. The cosine error was less than 10% for solar zenith angles less than 60°. The response uniformity along the 1 meter sensing length varies less than +/- 7%. Ground based sensing was conducted using spectrometer and quantum sensor to correlate with image spectral reflectance and photo synthetically active radiations (PAR) to estimate dry matter content using the following expressions (Baret and Guyot 1991) :

$$DM = \varepsilon \sum f_{ipar} I_0 dt \quad (1)$$

DM is the dry matter ($g.m^{-2}$), f_{ipar} fraction of intercepted photo synthetically active radiation (400-700 nm)(ipar) I_0= photo synthetically active radiation (par) incident above the canopy ($MJ.m^{-2}.day^{-1}$); ε is the dry matter radiation quotient ($g.MJ^{-1}$). The photo synthetically active radiation can be expressed as:

$$I_o = 0.5Q \tag{2}$$

$$APAR = (I_o + R_S) - (T_c + R_c) \tag{3}$$

$$IPAR = I_o - T_c \tag{4}$$

$$f_{apar} = \frac{APAR}{I_o} \tag{5}$$

$$f_{ipar} = \frac{IPAR}{I_o} \tag{6}$$

APAR: Absorbed photo synthetically active radiation ($\mu mol/m^2/s$), IPAR: Intercepted PAR, *fipar:* Fraction of intercepted PAR. The dry matter (DM) can be estimated from the following expressions:

$$DM = \varepsilon \sum f_{ipar} I'_o \Delta t \tag{7}$$

$$I'_o = 0.5Q \tag{8}$$

$$I_0' = 0.00833 I_o E \tag{9}$$

$$DM = \varepsilon \sum_{i=1}^{n} f_{ipar_i} I'_{o_i} \Delta t_i \tag{10}$$

$$= \varepsilon [f_{ipar_1} I'_{o_1} \Delta t_1 + f_{ipar_2} I'_{o_2} \Delta t_2 + \ldots\ldots\ldots\ldots + f_{ipar_n} I'_{o_n} \Delta t_n]$$

Where $\Delta t_i = t_i - t_{i-1}$, DM: Dry matter biomass (g/m^2), ε: Dry matter radiation quotient (1.06-2.53 g/MJ), t: Time Period, Q: Global Solar Radiation, ($MJ/m^2/day$), E: Photons Energy, 1 ev=$1.602x10^{-17}$ J, 1$\mu mol/m^2/s$=$6.02x10^{17}$ Phontons/s/m^2.

2.2.1. Reflection

A spectrometer (QC, CID, Inc, Ocean Optics) was used to measure the reflectance from the crop canopy (Figure 1-e, f). The reflection is expressed as percentage relative to the reflection from a standard reference substance

$$\%R = \frac{S_\lambda - D_\lambda}{R_\lambda - D_\lambda} \times 100\% \tag{11}$$

A lambertian reference white surface was used for calibration of the spectrometer sensors and Spectrasuite (Ocean Optics) software was used to get the spectral signature and data acquisition from canopy.

2.2.2. Normalized Difference Vegetation Index (NDVI)

The Normalized Difference Vegetation Index (NDVI) is the most commonly used vegetation index. It is calculated by dividing the difference of NIR and red reflectance by the summation of NIR and red reflectance. The NDVI value ranges from –1 to 1. If there is no vegetation, the NDVI value is equal to 0. If the NDVI value is close to 1, it indicates dense vegetation. The NDVI can be expressed as:

$$NDVI = \frac{(NIR - R)}{(NIR + R)} \tag{12}$$

2.2.3. Green Normalized Difference Vegetation Index (GNDVI)

The GNDVI proves to be more useful for assessing canopy variation in green crop biomass. GNDVI can be expressed as (Huete 1988, Huete and Liu 1994):

$$GNDVI = \frac{NIR - G}{NIR + G} \tag{13}$$

where: G – green spectral band; R – red spectral band; RE – red-edge spectral band; NIR – near infrared spectral band. To represent each of these spectral bands, the reflectance was averaged in the following wavelength range: G – 549.04 to 551.77 nm (central wavelength is 550 nm); R – 669.01 to 671.72 nm (central wavelength is 670 nm); RE1 – 699.08 to 701.66 (central wavelength is 700 nm) RE2 – 749.26 to 751.78 nm (central wavelength is 750 nm); NIR – 799 to 801.8 nm (central wavelength is 800 nm)

2.3. CIR and RGB Image Interpretation

The layout of the field is depicted (Figure 2). A lab view based real time algorithm was developed to capture images from the field over the growing seasons.

Initially, 20 preset positions were set to cover the each of the fields. The 50 mm fixed focal length was chosen to capture images. The NIR, red and green channel were averaged in the image acquisition process.

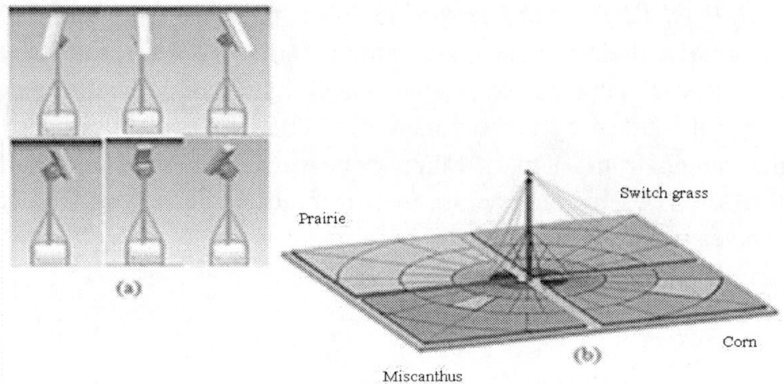

Figure 2. (a) Image acquisition concept using multispectral camera; (b) Field plots distribution and tower location at the energy farm, University of Illinois at Urbana-Champaign campus.

The tower coordinates and the ground reference points were surveyed using an RTK GPS unit. The stand-alone images for the reference points present the crop response and physiological changes. The ground reference points for Miscanthus (M1, M2, M3 and M4) were observed during the growing season and data was collected from September to December 2009.

The sensing system was established during August and images were ready for acquisition from September. The ground control points for mixed prairie grass (P1, P2, P3 and P4) were placed inside an 8 m by 8 m plot to track the vegetative responses. The reference points for switch grass (S1, S2, S3 and S4) were marked inside the field. The field spectrometry was limited to three energy grasses, however, for geo referencing the corner points for four fields: corn, Miscanthus, switchgrass and prairie grass were surveyed.

2.4. Development of On-the-Go Sensing System from Data Acquisition Vehicle (DAV)

The Data Acquisition Vehicle (DAV) was proposed as on-the-go close measurement for bioenergy crops over the growing seasons (Xiong et al. 2011). All the drive modules, steering modules, and lifting leg units, and the chassis frame and operating arm were assembled (Figure 3-a).

In regard to the reconfigurability of DAV, the basic kinematical dimensions of the vehicle were needed to design. The dimensions could be increased upto certain limit when fully extended through the adjustable mechanism.

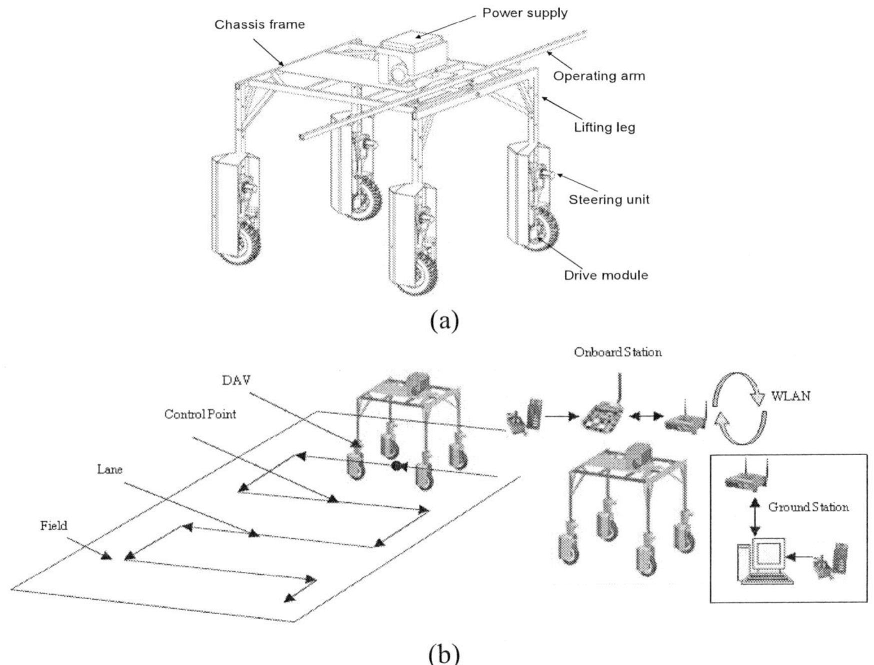

Figure 3. Data Acquisition Vehicle Control System (a) Overall view, (b) Sensors and path-way design.

In regard to the reconfigurability of DAV, the basic kinematical dimensions of the vehicle were needed to design. The dimensions could be increased upto certain limit when fully extended through the adjustable mechanism.

The design also allowed the add-on of sensors, actuators, and necessary monitoring devices, supporting up to a 300-kg payload (Figure 3-b). The field test was carried out after the control system was developed.

2.5. Development of on-the-Go Sensing System using an Autonomous Tractor

The biomass plant height under the tree canopy could be measured from an autonomous tractor over the growing season (Figure 4). An URG-30 LX laser range finder was used for monitoring biomass plants height (Figure 5). The height for a single plant can be exactly measured by a measuring scale or laser range finder, but the crop height inside the field was difficult to define.

The mean height from the cloud scan points was considered to define the plants height. The ground installation height was measured at the beginning and then laser reflection height based on time of flight principal was recorded. Plant height (h_p) was obtained from the difference between installation height of laser to ground (h_g) and reflection heights (h_f) (Figure 5).

Figure 4. Concepts of monitoring plant growth using an autonomous tractor and laser range finder.

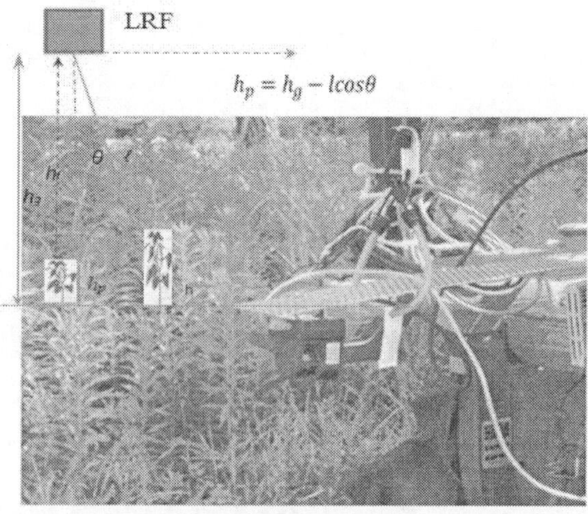

Figure 5. An URG laser range finders for measuring plant height from autonomous tractor.

3. Results and Discussion

3.1. Image Acquisition from Stand-Alone System

The CIR and RGB images were captured on 23 September, 3 November and 3 December 2009 from the stand-alone system for Miscanthus, switch grass and prairie (Figure 6).

Since the first set images were collected during 23 September, therefore the second sets of images were collected on 3 November and third sets of images were collected on 3 December. During September, the switch grass images reflected more NIR and absorbed red. The plant vigor and growths were good in these stages. The reference data was collected from the ground reference points and during this time period.

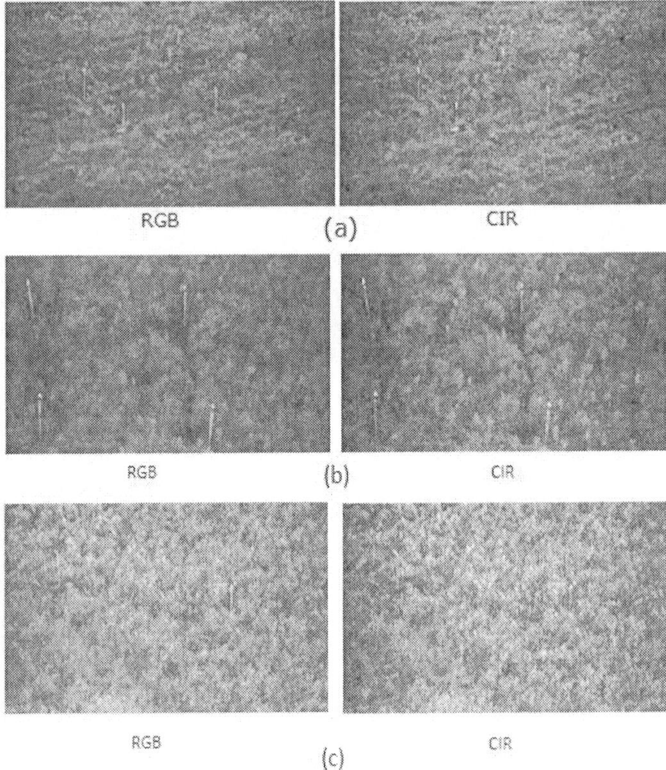

Figure 6. RGB and CIR images captured from stand-alone sensing system for Miscanthus, Switch grass and prairie field.

The NIR reflectance was decreased during November and December. The stand-alone images for the ground reference points present the crop response and physiological changes.

In the September, the crops were green, the NIR reflectance and Red band absorption were higher. The RGB images also represent the changes of canopies. The perennial crops are the first of their growth. The canopies would denser as the year increases of perennial growth.

3.2. Ground Sampling Data

3.2.1. Dry Matter Biomass

The dry matter biomass was estimated using photosyntically intercepted radiation and dry matter accumulation was calculated from the observation of September, November and December during these 70 days interval for Miscanthus, Switch grass and prairie grass (Figure 7).

The Miscanthus field was the first year of it's perennial growth and was not grown properly. On the other hand, switch grass and prairie were grown better than Miscanthus and biomass conversion from light interception was high.

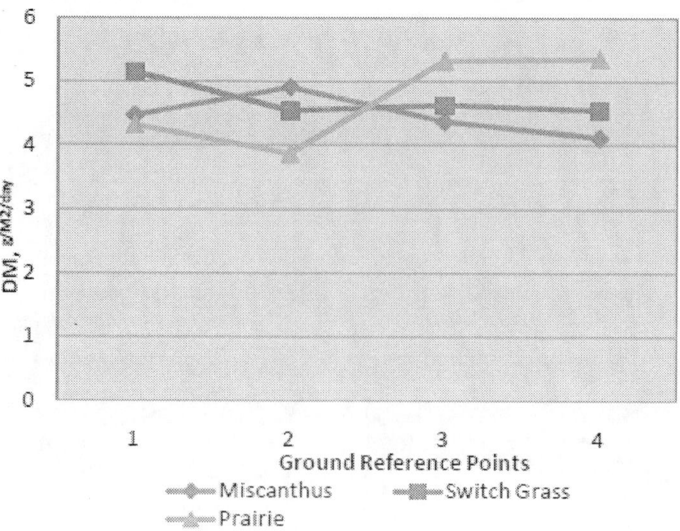

Figure 7. Dry Matter biomass estimation from Line Quantum Sensing Data using intercepted PAR radiation (September 23 to December 3).

3.2.2. Vegetative Indices

The images were captured from 38 m height during 12-3 pm. The temporal resolution was the major advantages for image database. In the, ground sampling, the reference points were selected to keep tracking the vegetation indices and intercepted solar radiation for the canopy. The spectrometer response from Miscanthus canopy was analyzed for NDVI and GNDVI (Figure 8-a).

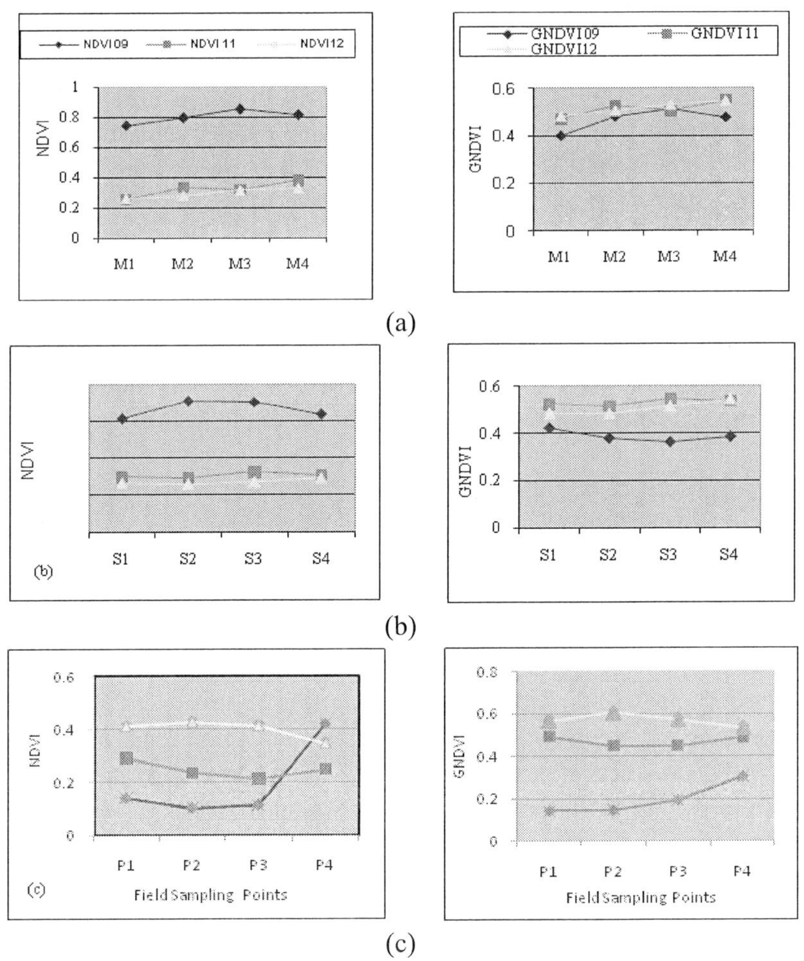

Figure 8. NDVI and GNDVI Trajectories at the ground reference points (a) Miscanthus, (b) Switch grass, and (c) Prairie.

NDVI are related with RED and NIR band and chlorophyll absorption. On the other hand the GNDVI was related with green band. CIR image on 23 September had more NIR information and gradually decreased during November and December.

The RGB images have also shown for comparison based on visible wavelengths. The GNDVI indices observations were closer during these months. The NDVI and GNDVI trajectories are depicted for switch grass during September, November and December (Figure 8-b). In September, the NDVI was higher than the November and December. Over the growing season NDVI varies more than GNDVI. On the other hand, the GNDVI was closer during September, November and December. In the prairie field, the NDVI curve for September at the point 4 had noises and did not represent the regular response (Figure 8-c). This could be occurred due to measurement error or irregular canopy structures.

3.2.3. Spectral Responses and NDVI Maps

The spectral responses and NDVI map of Miscanthus were analyzed for September, November and December (Figure 9). The NDVI maps show the spectral resolution from crop response based on NIR and red band information. The brighter part of the pixel indicates the healthy crops and grey or black indices the non-healthy crops. This depends on chlorophyll absorption which was closely associated with red band. The noises were increased during November and December. The chlorophyll absorptions were significantly reduced and NIR bands which had noises. There were couple of reasons involved; the reflectance was influenced using white Lambertian surface based on solar radiation. During September, the number of bright pixels was higher and reduced gradually during November and December. The NDVI map was required to calculate the yield estimation of crops. The statistical analysis from ground sampling per pixel helps to estimate the amount of biomass. The responses from switch grass canopies and NDVI maps were generated from CIR images helped to understand feedstock's quality over the growing season (Figure 10). For prairie, the spectral responses and NDVI maps were analyzed (Figure 11). The brightness indicates higher pixel value and the crops reflected more NIR. In December, NIR wavebands had noises.

3.3. Data Acquisition Vehicle (DAV)

3.3.1. Prototype Configuration

The DAV was developed with the ability to adjust its configuration to adapt to the different operating environment of the *Miscanthus* field. For the adjustable lifting mechanism, each leg had two square tubes sliding each other through the guide of two bushings.

The tubes were bolted for flexibility and easily adjustable application. The clearance could be increased from 3 to 4 m. Because *Miscanthus* grows very fast, the grass spreads out extensively and occupies a portion of intra-row room that should be followed by the DAV.

To avoid damaging crops that have expanded out to the intra-row, the wheel gauge of the DAV needs to be changeable, following the certain track of empty intra-row space. The drive modules, steering modules, lifting leg units, and chassis frame were assembled together (Figure 12). Particularly, the overall specifications: Locomotion 4WD4WS, length 3.3 m, width 3.2 m, height 3.1-4.1, weight 1500 kg, clearance 3-4 m, wheelbase 3.1 m, wheel gauge 2.85, 3, 3.15 m, speed 1 m/s, Maximum traction 2812 N, Maximum steering torque 130 N.m, Turning radius 0 m, Operating time 10 hr, power supply 6.5 kW.

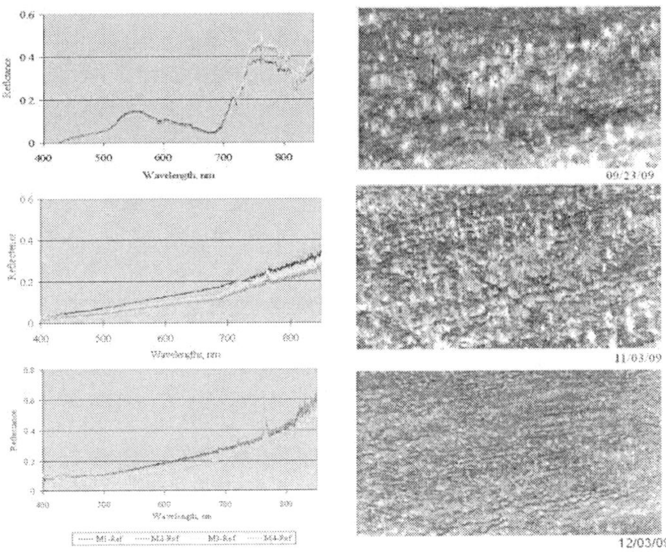

Figure 9. Reflectance from the Miscanthus during growing season on left side and NDVI map at the right side.

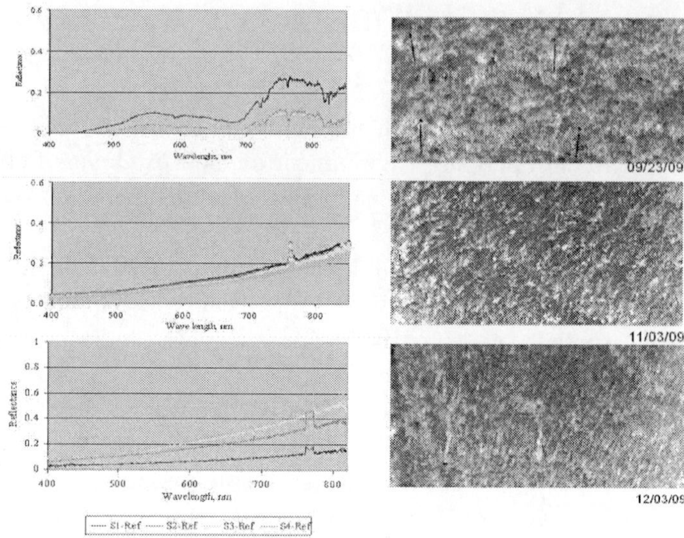

Figure 10. Reflectance from the switch grass during growing season on left side and NDVI map at right side.

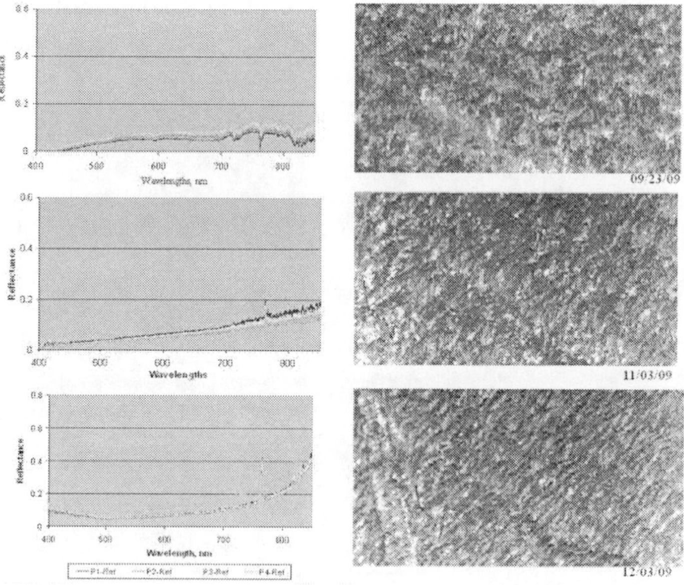

Figure 11. Reflectance from the prairie grass during growing season on left side and NDVI map at right side.

The DAV system was a mobile crop monitor and data collector. The remote sensors could be equipped on the DAV. A frame of remote sensing images can be collected which covers a small area around this point. DAV platform could be controlled either automatically or manually.

Figure 12. Prototype of on-the-go Data Acquisition Vehicle (DAV) at the University of Illinois at Urbana-Champaign.

The navigation path and control points can be stored as pre-set path and controled autonomously. The vehicle has been designed to operate from joystick either at ground station or onboard.

The field test was carried out after the control system was developed. Sensors on the DAV included cameras, transducers, tilt sensors, speed sensors, and angle sensors.

A stereo camera was at the front of each wheel axle looking forward and down to provide images of crop at inter-row and intra-row lines and to detect obstacles.

3.4. On the-go Autonomous Tractor-based Sensing

An actual size autonomous tractor was used in the experiments to estimate mean plants height under the tree canopy (Figure 13). The relative distance of the biomass plants and height were calculated from the cloud scan points of data. (Figure 14)

The laser range finder was used under the canopy biomass height estimation. The plant relative distance and reflection height was measured using an URG laser range finder (Figure 15).

Figure 13. Prototype of on-the-go sensing platform at the Agricultural and Forest Research Center, University of Tsukuba.

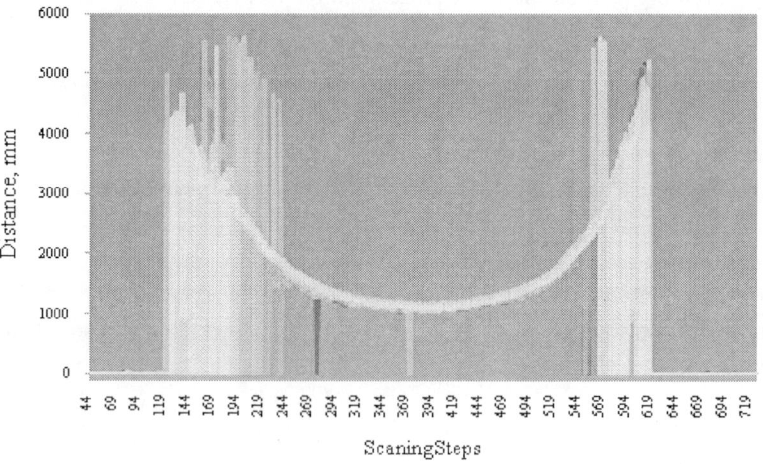

Figure 14. Distance from laser to plant surface using URG Laser (Step 44-716).

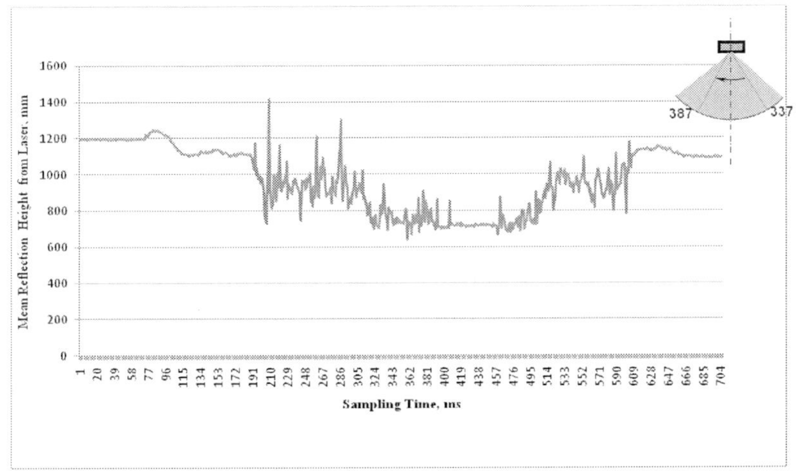

Figure 15. Mean relative distance (l) of plants from laser range finder. (Steps 337-387).

Conclusion

The instrumentations and data processing for ground-based remote sensing systems were developed to monitor bioenergy crops. Firstly, the stand-alone system was developed which was capable of collecting images over the growing seasons to enable site-specific managements. The system was independent compare to the conventional systems that depends on weather, flying opportunity and temporal resolutions. Especially satellite has extensive limitations on revisit time on the experimental area. The ground reference sensing was done to realize the spectrometer responses for crop growth and quantum sensing to estimate biomass accumulation from intercepted solar radiation for Miscanthus, switch grass and prairie grass.

The NDVI maps were developed to visualize the spectral signature and understand plant stress based on the near-infrared information. Secondly, on-the-go data acquisition vehicle was developed to overcome problem of close proximity sensing of Miscanthus. Thirdly, we were concerned about illumination problem for optical sensors under the tree canopy. In this regard, autonomous tractor-based instrumentation was developed for on-the-go measurement of biomass feedstock. The further experiments and real time processing of images and data from sensors will be transmitted through wireless communication to local server for sharing with interdisciplinary researchers.

Acknowledgments

The authors would like to thank Energy Biosciences Institute (EBI), University of Illinois at Urbana-Champaign to support the program "Engineering Solutions for Biomass Feedstock Production". The present research work is the part of this program and authors extended their gratitude's to Yuliang Zhang, Xiong Yonghua, Yanshui Jiang, Bin Zhao, Hx Liu, Francisco AC Pinto, and Casimiro DG Junior for their contribution in design and development of sensing systems at the University of Illinois at Urbana-Champaign. The authors also expressed their gratitude's to Agricultural and Forest Research Center (AFRC), University of Tsukuba to allow the facilities and instruments for conducting experiments.

References

Ahamed, Tian L., Takigawa, T., and Zhang Y. (2009). Development of Auto-Hitching Navigation System for Farm Implements using Laser Range Finder, *Transactions of the American Society of Agricultural and Biological Engineering (ASABE),* Vol. 52(5):1793-1803

Ahamed T., Kulmutiwat Supachai, Thanpattranon P., Tuntiwut S., Ryozo, N., and Takigawa, T. (2011). Monitoring of Plant Growth Using Laser Range Finder, *ASABE Paper* No.1111373. St. Joseph, Mich.

Bajwa, S. G., and L. Tian.(2001). Modeling and mapping of spatial weed density within a soybean field from aerial CIR images. *Transactions of the ASABE* 44(6): 1965-1974.

Baret, F., and Guyot, G.(1991). Potentials and limits of vegetation indices for LAI and APAR assessment, *Remote Sensing of Environment*, 35: 161-173.

Bausch, W. C., and H. R. Duke. (1996). Remote sensing of plant nitrogen status in corn. *Trans. ASAE* 39(5): 1869−1875.

Gitelson, A., Kaufman, Y. J. and Merzlyak. M. N. (1996). Use of a green channel in remote sensing of global vegetation from EOS-MODIS. *Remote Sens. Environ.* 58: 289-298.

Huete, A. R. (1988). Asoil-adjusted vegetation index (SAVI). *Remote Sensing of Environment* 25: 295-309.

Huete, A., and Liu H. Q. (1994). An error and sensitivity analysis of atmospheric- and soil-correcting variants of normalized difference vegetation index from MODIS-EOS. *IEEE Transactions on Geoscience and Remote Sensing* 32(4): 897-905.

Jensen, J. (2000). *Remote Sensing of Environment: An Earth Resource Perspective.* Englewood Cliffs, N.J.: Prentice Hall.

Jørgensena U, Morttensen J, and Ohlsson, C., (2003). Light interception and dry matter conversion efficiency of miscanthus genotypes estimated from spectral reflectance measurements, *New Phytologist* (2003):157: 263-270.

Lyon, J.G., D. Yuan, R.S. Lunetta, and C.D. Elvidge. (1998). A change detection experiment using vegetation indices. *Photogrammetric Engineering and Remote Sensing* 64(2): 143-150.

Milton EJ. (1987). Principles of field spectroscopy, *Int. J. Remote Sensing,* 8(12):1807-1827.

Thorp, K., and L. Tian. (2004). A review on remote sensing of weeds in agriculture. *Precision Agriculture* 5: 477-508.

Ting KC, Hansen, A, Zhang, Q., Grift, T., Tian, L., Eckhoff, S., and Rodriguez L. (2008). *Engineering Solutions for Biomass Feedstock Production,* http://www.energybiosciencesinstitute.org.

Tumbo SD, Salyani M, Whitney JD, Wheaton and TA, Miller WM. (2002). Investigation of laser and ultrasonic sensors for measurements of citrus canopy volume, *Applied Engineering in Agriculture* 18, 367-372.

Vargasa LA, Andersena, M.N., Jensenb, C.R., Jørgensena U. (2002). Estimation of leaf area index, light interception and biomass accumulation of Miscanthus sinensis 'Goliath' from radiation measurements, *Biomass and Bioenergy* 22: 1-14.

Xiang, H., and Tian, L.F. (2007) Artificial intelligence controller for automatic multispectral camera parameter. *Transaction of ASABE*, 50(5),1873-1881.

Xiong Y. Tian, L., Ahamed, T., and Zhao B. (2011). Development of the Reconfigurable Data Acquisition Vehicle for Bio-energy Crop Sensing and Management, *Journal of Mechanical Design*, 134, 015001-7.

Yao, H. 2004. Hyperspectral imagery for precision agriculture. PhD dissertation, Champaign, IL: University of Illinois at Urbana-Champaign, *Department of Agricultural and Biological Engineering*.

Yao, H., and Tian, L. (2004). Practical methods for geometric distortion correction of areal hyperspectral imagery. *Applied Engineering for Agriculture,* 2004, 20(3):367-375.

Yang, C, Everitt, J.H., and Bradford, J.M. (2002). Airborne hyperspectral imaging and yield monitoring of grain sorghum yield variability. *ASAE Paper No.* 021079. St. Joseph, Mich.: ASAE, 2002.

In: Remote Sensing
Editor: Enner Alcântara

ISBN: 978-1-62417-140-6

Chapter 3

REMOTE SENSING AS A SUPPORT TOOL FOR WATER RESOURCES MANAGEMENT IN SUB-SAHARAN AFRICA: A REVIEW

***W. Gumindoga*[1,*]*, A. T. Kabo-bah*[2] *and M. D. Shekede*[3]**

[1]University of Zimbabwe, Dept. of Civil Engineering, Harare, Zimbabwe
[2]Hohai University, Nanjing, China
[3]University of Zimbabwe, Dept .of Geography and Environmental Science, Harare, Zimbabwe

ABSTRACT

This chapter provides an insight into how remote sensing technology can provide near real time answers to hydrological modelling, flood forecasting, water conservation and efficiency in irrigation and development focusing on Sub-Saharan Africa. It can be stated that the effective use of remote sensing is parallel to better management practices of water and possibly a revolutionary contribution towards answers to the MDGs and current talks at the COPs for the most part of Africa. In recent past, globally, Remote Sensing has demonstrated consideration for use in rainfall estimation, snow and glacier studies, irrigation water management, reservoir sedimentation, watershed management, disaster management, water quality assessment, wetland mapping, and ground water assessment. The challenge however for scientists is the fast growth

* E-mail: wgumindoga@gmail.com.

of remote sensing missions, working with new data structures and hardware technology deficiency in handling large amounts of data in relatively shorter time steps. Generally, the temporal resolution of most satellites have improved (e.g. MSG temporal resolution is 15mins) and this is a huge opportunity to support studies in water science. Therefore, to complement the huge efforts of this new remote sensing, this work provides an overview of state-of-the-art applications specific to water resource management towards the improvement of agricultural growth and subsequently safeguarding food security with emphasis on the Sub-Saharan Africa. The chapter identifies real problems that remote Sensing technology has resolved and some of the challenges this field needs to be addressed to ensure that it achieves the most impact in future. In addition, several case studies are presented to demonstrate the utility of remote sensing in water conservation for water supply for agricultural purposes, The work envisages policy makers, planners and water resources managers to appreciate the potential of remote sensing in the management of fragile and limited water, resources that are critical for infrastructure development in Africa and other the developing countries.

1. INTRODUCTION

Water is a ctitical renewable and finite natural resource that supports human life through agriculture, industry and domestic uses. The environmentally sustainable development and management of water resources is a critical and complex issue for both rich and poor countries (World Bank, 2003). In fact, the security and sustainable development of water resources is one of the key problems of the 21st century. Improved water management can make a significant contribution to achieving the Millennium Development Goals (MDGs) and discussions on Climate Change, Mitigation and Adaptation Measures. As freshwater becomes an increasingly scarce resource, better methods for managing water resources must be adopted.. For these water resource managemement methods to be effective, they should allow for timely and accurate data collection. In this regard, remote sensing has been identified as key to solving water resource management across the globe. Indeed, current international initiatives such as the Global Earth Observation System of Systems 10-Year Implementation Plan have identified remote sensing (RS) as the key to helping to solve the world's water problems.

The growing international gatherings such as the American Society for Photogrammetry and Remote Sensing (ASPRS) conference, Land Surveying Summit, Geospatial World Forum, International Society for Photogrammetry and Remote Sensing (ISPRS) Congress, AfricaGIS conference, GIScience, Open Source Geospatial Research and Education Symposium (OGRS), African Association of Remote Sensing of the environment (AARSE) Conference, International Cartographic Conference and many more, indicate the overwhelming awareness of the need to use remote sensing for improving the life of mankind.

The use of RS technologies have been and can be useful in providing a cost-effective means of replacing or complimenting field data collection especially in Sub-Saharan Africa where data is scarce. In particular, RS data provides coverage over large and remote areas with systematic, repetitive data captures and can also be integrated with field or remotely collected data (i.e. real-time) to produce effective up-to-date predictive and analytical products. Specifically, RS is important in providing near real time data required in hydrological modelling, flood forecasting, water conservation and efficiency in irrigation use and development.

The effective use of RS in water resources management provides better management practices of water resourcers and possibly a revolutionary contribution towards answers to the MDGs and current talks at the COPs for most part of Africa. Several studies across the globe have demonstrated the utility of RS in rainfall estimation (Hong et al 2004), snow and glacier studies (Rees, 2006) irrigation water management (Kim et al, 2008), reservoir sedimentation (Ritchie and Cooper 1998) watershed management (Bastiaanssen,1998), disaster management (Inglada and Giros 2004), water quality assessment Chen et al 2007), wetland mapping (Ozesmi and Bauer 2002) and ground water assessment (Saraf and Choudhary 1998). The challenge however for scientists is the fast growth of remote sensing missions, working with new data structures and hardware technology deficiency in handling large amounts of data in relatively shorter time steps. Generally, the temporal resolution of most satellites have improved (e.g. MSG temporal resolution is 15mins) and this is a huge opportunity to support studies in water science.

Therefore, to complement the huge efforts of this new dispensation in RS, this paper provides an overview of state-of-the-art applications specific to water resources management towards the improvement of agricultural growth and subsequently safeguarding food security with emphasis on Sub-Saharan Africa. The work seeks to identify real problems that the RS technology has helped to resolve and some of the challenges this field needs to be addressed in the future.

The authors are aware of the fact that the review here underestimates the hidden works undertaken by most research departments and scientists in this particular area. Hence, the review provided here is only a snapshot collection and to emphasise the fact that indeed, remote sensing has some practical benefits for effective water resources management in sub-Saharan Africa. In view of this, related case studies that have demonstrated the useof RS for conserving water resources as a means for improving water supply for agricultural purposes have been included.

The work seeks to show policy makers, planners and water resources managers the potential of RS in managing the limted and fragile water resources that are critical for agriculture, industry and human wellbeing in Africa and other developing countries.

In addition, water related problems in Africa could be better understood and tackled using the combined concepts and principles of Integrated Water Resources Management (IWRM), hydrological and water quality modeling and Geographic Information Systems (GIS) and Remote sensing (RS) tools. Adopting the IWRM concept could make a significant contribution to achieving most of the Millennium Development Goals (MDGs), since water relates to health (malaria, diarrhoea), food (supplementary irrigation, home gardens, fish water) and ecosystems integrity.

Key Terminologies

It is very important to define keys term such as Remote Sensing, Earth Observation, Hydrology, Integrated Water resources management (IWRM) and hydrological analysis. There are various explanations and definitions of these terms over the years as a result of the growth in understanding of sustainable development and the need to encompass climate change issues in our future planning of water resources.

Remote Sensing and Earth Observation

Traditionally, the field of remote sensing started with land surveying and later evolved into aerial surveying as a result of man's ability to view landscapes using helicopters. Over time, this changed with the advent of satellite technology since the first launch of Landsat in 1972. Thus several definitions have been proposed. For instance, *Remote Sensing* has been defined as the science of acquiring, processing and interpreting images that record the interaction between electromagnetic energy and matter (Sabins 2007).

Another popular definition is by Lillessand and Kiefer (1994) who define *Remote Sensing* as the science and art of obtaining information about an object, area, or phenomenon through the analysis of data acquired by a device that is not in contact with the object, area, or phenomenon under investigation. Buiten and Clevers (1993) define *Remote Sensing* as the instrumentation, techniques and methods to observe the earth's surface at a distance and to interpret the images or numerical values obtained in order to acquire meaningful information of particular objects on earth. Somehow, through last decade, there has been the argument among scientists about a more general term to encompass all aspects of science capturing information about the earth.

Though not clearly specified under a particular reference, it is now broadly accepted that *earth observation* is used to define all aspects of data acquisition from earth with or without physical contact with the object or natural phenomena under question, the data analysis and visualisation of information in either via the world wide web, mobile phones or just on the desktop. This is because, in recent times, the combined interactive use of in-situ data and remote sensing data is synonymous to the validating of data and results from remote sensing. Therefore, earth observation would mean the combined use of remote sensing data and in-situ data to better understand the earth fluxes using all appropriate and possible visualisation gadgets present at the time.

Hydrology

This is the science of water (Brutsaert, 2005) that deals with those aspects of the cycling of water in the natural environment that includes the following:

- the continental water processes, i.e. the physical and chemical processes along the various pathways of continental water (solid, liquid and vapor) at all scales, including those biological processes influencing this water cycle directly
- the global water balance, i.e. the spatial and temporal features of the water transfer (solid, liquid and vapor) between all components of the global system – atmosphere, oceans and continents, in addition to stored water quantities and residence times in these compartments

Hydrological Analysis

On the other hand *Hydrological Analysis* is the determination of the amount and/or rate of water that will be found at a given location and at a given time under natural condition.

Hydrology deals with, in a quantitative manner (via a mathematical model), the continental water processes and the global water balance - the water and energy processes on earth – in order to be able to predict the changes in future. Accounting for amounts of water entering and leaving the unit of observation is based on the mass conservation law:

$$Q_{in} - Q_{out} = \Delta V \tag{1}$$

where: Q_{in} is the inflow, Q_{out} is the outflow and ΔV is the change of storage.

Integrated Water Resources Management

IWRM is the integrating concept for a number of water sub-sectors such as hydropower, water supply and sanitation, irrigation and drainage, and environment. An Integrated Water Resources perspective ensures that social, economic, environmental and technical dimensions are taken into account in the management and development of water resources. Although the science of hydrology is an essential basis for the practice of water resources management, water resources decisions are ultimately determined by political, legal, economic and social factors (Kusangaya, 2007). Integrated Water Resources Management is therefore the process by which governments, the private sector, and/or individuals reach and implement decisions that are intended to affect the future availability and/or quality of water for beneficial uses or the risk of water-related hazards that threaten beneficial activities. For many countries in Africa, the process of Water Resources Managementinvolves:

- Assessment of the present and future *quantity* of water available from surface and/or groundwater resources
- Assessment of the present and future *quality* of surface and/or ground water
- Assessment of the present and future *frequencies* with which human activities will be subject to water-related hazards such as flooding, drought and erosion of various magnitudes.

- Equitable sharing of water resources and use to avoid conflicts in river basins
- Sustainable use of the water resources to ensure that future generations benefit from the resource

2. Remote Sensing Applications in Water Resources Management

This section gives a snapshot description, analysis and evaluation of the various efforts that have been made in the application of remote sensing in water resources management in Sub-Saharan Africa. However, it is impossible to present in perspective all the applications of remote sensing for water resources management. Therefore, focus is given to the application of remote sensing in the quantification of the following water cycle components:

a. Digital Elevation Models (DEMs) for Watersheds
b. Rainfall Retrieval and analysis
c. Evaporation from land and open water
d. Water in soils
e. Rainfall-runoff processes
f. Water quality modelling
g. Monitoring water levels and flood extents in river basins

Digital Elevation Models (DEMs) for Watersheds

Introduction

Terrain patterns play an important role in determining the nature of water resources and related hydrological modelling. DEMs, offering an efficient way to represent ground surface, allow automated direct extraction of hydrological features (Garbrecht and Martz, 1999), thus bringing advantages in terms of processing efficiency, cost effectiveness, and accuracy assessment, compared with traditional methods based on topographic maps, field surveys, or photographic interpretations.

However, researchers have found that DEM quality and resolution affect the accuracy of any extracted hydrological features (Quinn et al., 1995). Therefore, DEM quality and resolution must be specified according to the nature and application of the hydrological features (Kenward et al., 2000). Light Detection and Ranging (LiDAR) data is used to support a series of salinity and water management projects. The DEM derived from the LiDAR data has a vertical accuracy of 0.5m and a horizontal accuracy of 1.5m. Thus a high quality DEM can be used to derive much detailed terrain and hydrological attributes with higher accuracy. Table 1 provides a summary of some attributes some insight into the range of possible uses of DEMs in environmental modeling.

Table 1. Primary attributes derived from DEMs and their applications (adapted from Huggett, 2003)

ATTRIBUTE	APPLICATION
Altitude	Climatic variables (temperature, pressure); Vegetation and soil patterns; Potential energy determination.
Slope	Overland and subsurface flow; Erosion.
Aspect	Solar irradiance; Evapotranspiration.
Curvature	Flow acceleration; Erosion and deposition patterns and rates; Soil and land evaluation indices.
Drainage basin slope, area, and length	Time of concentration; Contributing volumes to runoff; Runoff attenuation times and velocities.
Stream lengths and slopes	Channel flow rates and velocities; Erosion rates and sediment yields.

Effective Hydro-Processing of a DEM

This is a cost efficient alternative to acquire terrain based information. Next to existing topographic maps, a DEM can be obtained from free archives (for example the Shuttle Radar Terrain Mission (SRTM) or extracted from RS sources. The increase in computer processing speed development of algorithms that handleDEM as well as the full integration of these algorithms in GIS softwares such as the Integrated Land and Water Information System (ILWIS) have provided hydrologists and water resource managers with hydroprocessing capabilities that were not previously available. DEM Hydro processing tools allow efficient processing of elevation information to extract a multitude of hydrologically relevant parameters.

The DEM Hydro processing tools are based on the D-8 flow direction model. In order to perform distributed or even lumped rainfall- runoff modeling, a multitude of information is needed (Beven and Kirkby, 1993). Part of this model input can now be provided through processing and analysis of a Digital Elevation Model (DEM) in combination with information extracted from other remotely sensed images of a selected model area. The development of DEM processing algorithms as well as relevant software to extract hydrologic information from DEM is increasing and makes it widely applied.

Case Study 1: DEM Hydro-Processing for the Upper Manyame Basin of Zimbabwe and Densu River Basin in the Ghana

GIS and RS based DEM hydroprocessing prepares input data that can be used for distributed or semi-distributed rainfall-runoff models such HEC-HMS, TOPMODEL HBV or MIKE SHE. As an example, a readily available Advanced Space born Thermal Emission Radiometer (ASTER) DEM with a near global coverage and a spatial resolution of of 30m was used to determine the hydrological parameters of the Upper Manyame subcatchment of Zimbabwe. The DEM Hydro-processing module found in the ILWIS GIS was

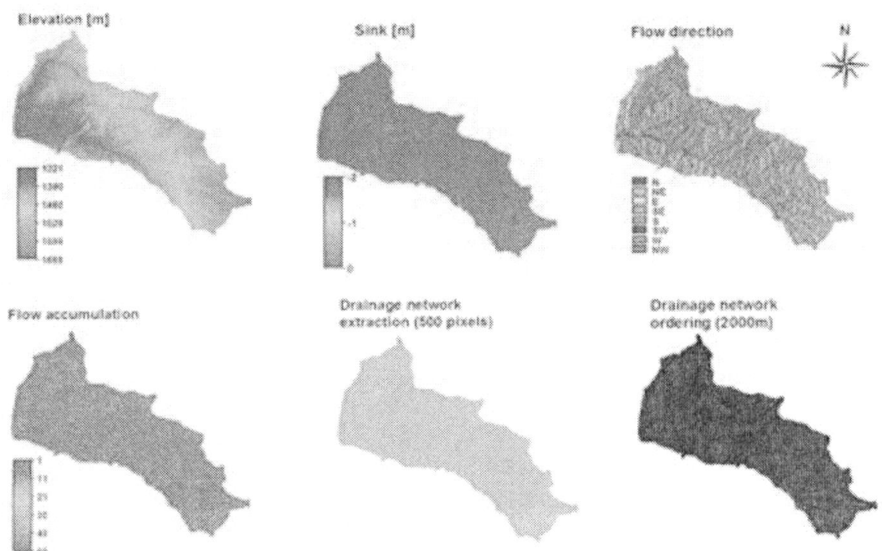

Figure 1. DEM Hydroprocessing results showing elevation, sinks, low direction, accumulation, drainage network.

used to extract the important parameters (see Figure 1). These are the original DEM, extracted sinks, flow direction, flow accumulation, drainage network extraction and drainage network ordering.

A similar process was used for the extraction of Densu River basin in Ghana using an SRTM 90m DEM. In this particular case study, it was realised that the derived basin characteristics were closely related to the original basin derived from topography maps (seeFigure 2).

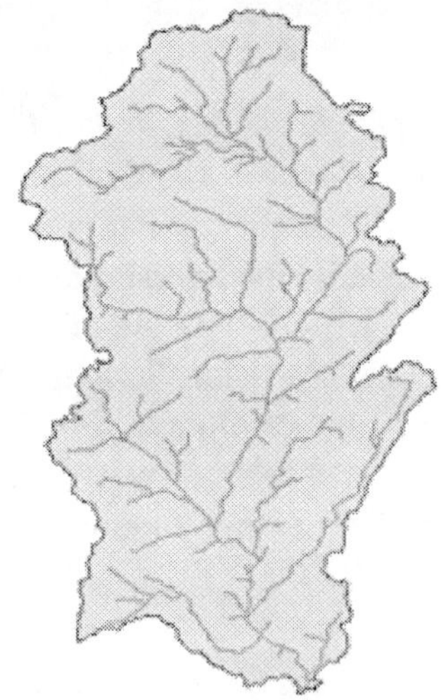

Figure 2. Densu river basin derived from SRTM 90m DEM in ILWIS Open. Source: Anornu et al. (2012).

Rainfall Retrieval and Analysis

Introduction

Accurate prediction of floods and drought is important for minimizing these natural hazards.For improved forecast of floods and droughts, accurate and timely estimates of rainfall are needed. Given that most parts of Africa have sparse and uneven distribution network of weather stations, estimation of rainfall in areas without is often unreliable or incomplete for the few available weather stations. The use of satellite-based weather data overcomes some of these problems. Thi particularly important if we consider the synoptic abilities as well as the regular and repetative coverage of nearly all of the earth's surfaces (Johnson et al. 1993). Such satellite-based rainfall data have the potential to be fed into hydrological models and thus be used to predict flooding and drought in Africa(Blackmore et al., 2009). The recent growing availability of earth observation (EO) products for precipitation offers great opportunity for improving hydrological modelling in Africa. Therefore, the presence of EO rainfall products could further provide more insights into river basin water budgets and catchment modelling.

It is however not possible to measure precipitation from a satellite directly.Rather the electromagnetic radiation is what is measured and rainfall estimates are then inferred from electromanetic radiation through algorithms. Several different types of satellite rainfall estimation algorithms exist. The main types are:

a. Algorithms that primarily use infra-red (IR) data
b. Algorithms that primarily use passive microwave (PM) data
c. Algorithms that use a combination of IR and PM data

The Tropical Rainfall Measuring Mission (TRMM) widely used for rainfall estimation for Africa is a visible and infrared scanning radiometer with a passive microwave imager and active microwave precipitation radar. Its altitude is 350 km with a non-sun synchronous orbit of 35 degrees. TRMM is a skewed orbit with a period of 91.5 minutes. Its approximately 15 orbits/day and in 2001, it was raised to 402 km. Figure 3 shows the diagrammatic view of the TRMM orbits.

Figure 3. TRMM orbits. Source NASA

Many algorithms also assimilate available real-time gauge data, and some now utilise data from the precipitation radar (PR) onboard the TRMM satellite. The University of Reading's TAMSAT (Tropical Applications of Meteorology using SATellite data) algorithm, which uses data from a single thermal infrared Meteosat channel, integrates satellite data and ground-based observations to derive information on rainfall over the African region.

Though this data is available on ten-daily (decadal), monthly and seasonal rainfall estimates, it has not been widely used as a tool to inform decision makers and planners, particularly in flood forecasting and agricultural sectors where information about the state of the rainy season is critical.

Table 2 gives a comparison of the IR/VIS and MW channels for rainfall estimation.

Table 2. Comparison of IR/VIS and MW channels

	IR/VIS	MW
Physical Robustness:	IR/VIS data reflect cloud-top conditions only and thus are more weakly related to actual rainfall rates over a wider range of conditions than MW radiances	Microwave radiances are sensitive to moisture throughout the cloud;
Space/Time Resolution:	IR/VIS data are available at 3 km resolution (MSG) on geostationary platforms, allowing looks in many locations every 15 minutes—suitable for extreme precipitation events at short time scales	MW instruments are presently restricted to polar-orbiting platforms, limiting views to 2 per day per satellite-more suitable for larger scales in time and space.

Case Study 2: TAMSAT Rainfall Estimates in the Volta Basin of West Africa

In this case study, Kabo-bah et al (2012) compared the satellite derived rainfall data and information to selected rain gauge station measurements highlighting the benefits and potential uses for satellite derived data especially in the Volta Basin where existing gauging stations have either been destroyed or are not in service due to poor maintenance over the years. The Volta Basin is an important river basin shared by six countries (Ghana, Burkina Faso, Ivory Coast, Mali, Benin and Togo).

The satellite derived data provide good geo-spatial representation of rainfall and useful information that can be utilized to facilitate hydrological modelling, flood forecasting, as well as planning and scheduling in irrigation schemes. Basically, the TAMSAT data were obtained through the courtesy of the GEONETCAST server at ITC (www.itc.nl). Rain gauge records were downloaded from (http://www.tutiempo.net/en/) for the same period.

Monthly cumulative rainfall totals were computed from the rain gauge readings. Examples of the interpolated local station measurements and TAMSAT derived rainfall are illustrated in Figure 4.

The results show that though the data sets vary in terms of absolute rainfall, their general geo-spatial patterns within the basin of interest are comparable. The authors observed that the undergoing improvements of TAMSAT algorithm to incorporate the use of Artificial Neural Networks (ANN) in Numerical Weather Prediction (NWP) model information for estimation processes could improve the results better.

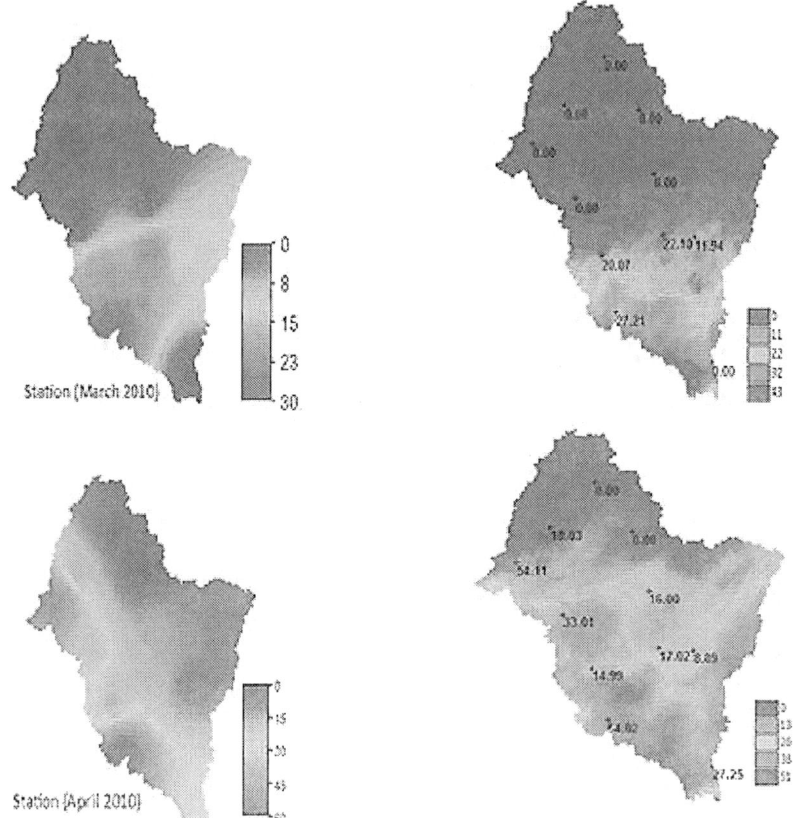

Figure 4. Examples of comparison of TAMSAT derived rainfall and in-situ measurements in the Volta Basin of West Africa.

Evaporation and Evapotranspiration Estimation

Introduction

Evaporation is the amount of water evaporated per m^2 surface (kg $m^{-2}s^{-1}$ or mmd^{-1}). Evaporation heat λ is the energy needed to evaporate 1 kg of water (J

kg^{-1}). Latent heat flux is the energy needed for evaporation (J $m^{-2}s^{-1}$ = Wm^{-2}). When measuring evapotranspiration, it should be clear that one is measuring water fluxes / water balance and energy fluxes / energy balance. This is measured both on the ground and from space. However measurements from space are indirect and until now only *energy balance* is used.

Energy Balance

Energy from the sun used for different 'purposes' one of them latent heat flux. Figure 5 shows the components of the energy balance which states that the Net radiation (Rn) consists of the sum of the latent heat flux (λE), the Sensible heat flux (H), the soil heat flux (G) and the chemical reactions (C).

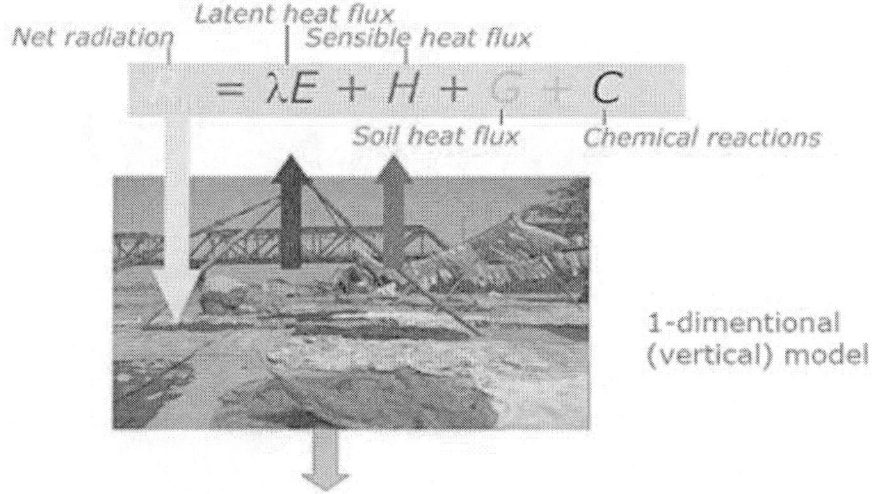

Figure 5. The Energy Balance terms.

Measuring the Energy Balance from Space

Cuurently direct measurement of energy balance from space is not possible. However, we can infer the amount of energy balance through quantification of terms of the energy balance from space.. Therefore we estimate *Rn*, *H* and *G*. Estimate $\lambda E = Rn\text{-}H\text{-}G$ using parts of the reflected spectrum and modelling, but how? The parameters of interest are in the *VIS – NIR – SWIR* such as:

i. Leaf Area Index
ii. Leaf Inclination Distribution
iii. Leaf water content
iv. Leaf chlorophyll content
v. Fractional Vegetation Cover
vi. Vegetation Height
vii. Absorbed Photosynthetically Active Radiation (aPAR)
viii. Reflected shortwave radiation
ix. Albedo

Case Study 3: The Surface Energy Balance System (SEBS) for Estimation of Turbulent Heat Fluxes and Evapotranspiration in the Zambezi Basin

The Surface Energy Balance System (SEBS) is briefly introduced. SEBS is scale invariant, so that it can be applied easily to different scales. Data of high or low spatial resolution from all sensors in the visible, near-infrared and thermal infrared frequency ranges can be used in the system (see Figure 6).

Based on a set of case studies, SEBS has proven to be capable to estimate turbulent heat fluxes and evaporation from point to continental scale with acceptable accuracy for low vegetation.

The results shown demonstrate that SEBS algorithm can be used for spatial temporal estimation of actual evaporation with an acceptable accuracy. Example of the use of SEBS algorithm in the Zambezi basin based on Kabo-bah et al (2011) is shown in Figures 7 and 8. The results indicated comparable measurements with ground estimates of evapotranspiration.

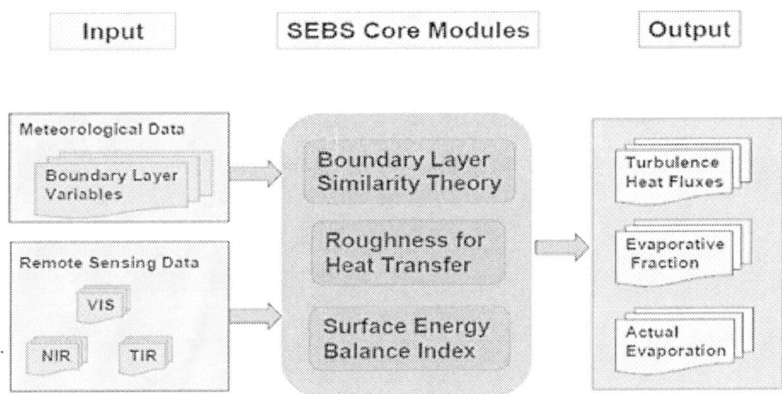

Figure 6. The SEBS structure.

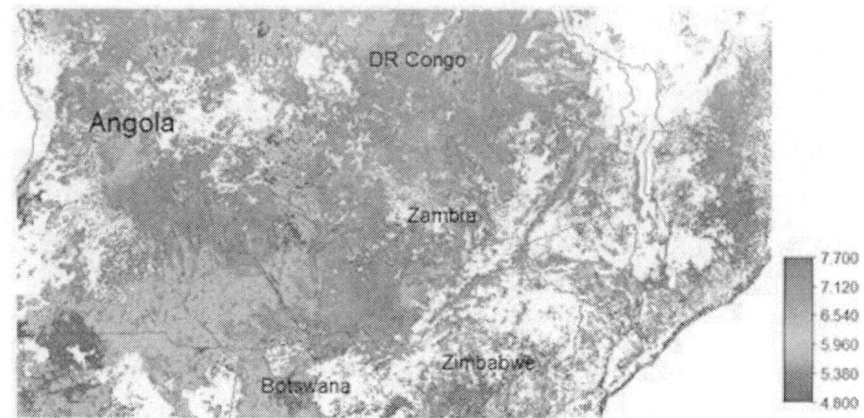

Figure 7. ET estimates on January 11, 2009 (measurements in mm/day).

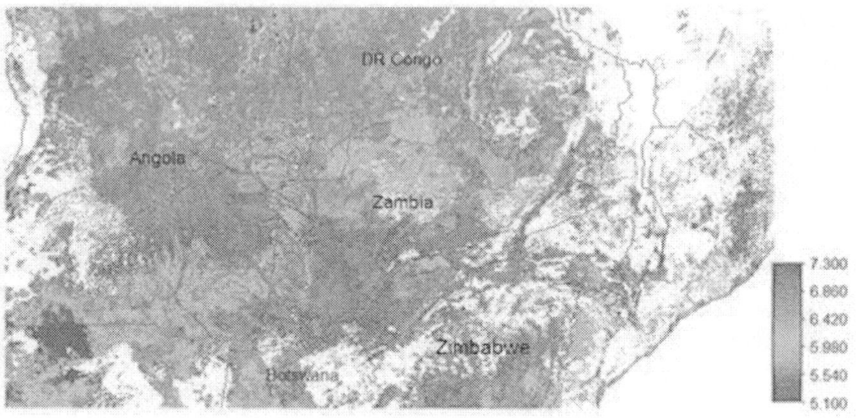

Figure 8. ET estimates on March 21, 2009 (Measurements in mm/day).

Water in Soils

Introduction

A distinction is made between the passive and active microwave imaging system used to measure soil moisture. The measurement principles are that radars measure the energy scattered back from the surface.

The radiometers measure the self-emission of the Earth's surface. Passive and active methods are interrelated through Kirchhoff's law:

$$Emissivity = 1 - \text{Re}\,flectivity$$

Table 3 shows the distinction between the passive and active sensors after the lecture notes of Wagner and de Jeu (2009).

Table 3. A distinction between the passive and active sensors (Wagner and de Jeu, 2009)

PASSIVE SENSORS	ACTIVE SENSORS
Detect the reflected or emitted electromagnetic radiation from natural sources Non-Imaging (ex. Microwave radiometer, magnetic sensor) Imaging (ex: cameras, optical mechanical scanner, spectrometer, microwave radiometer)	Detect reflected responses from objects irradiated by artificially-generated energy sources Non-Imaging (ex. Microwave altimeter, laser) Imaging (Real Aperture Radar, Synthetic Aperture Radar) *Radar*: Radio Detection and ranging (e.g. to detect airplanes, rain) *SAR*: Synthetic Aperture Radar, airborne systems developed in 1950s

Microwave Sensors and their application in Soil Moisture Retrieval

i. Microwaves (electromagnetic waves with wavelengths from1 mm – 1 m)
ii. All-weather, day-round measurement capability
iii. Very sensitive to soil water content below relaxation frequency of water (< 10 GHz)
iv. Penetrate vegetation and soil to some extent
v. Penetration depth increases with wavelength

Active Microwave Instruments

Both Synthetic Aperture Radar (SAR) and Scatterometer are active instruments on board satellites that measure the backscattering signal (Wagner et al., 1999).

SAR

- High spatial resolution
- Compatible with land surface variability

- Very low time sampling of a given region

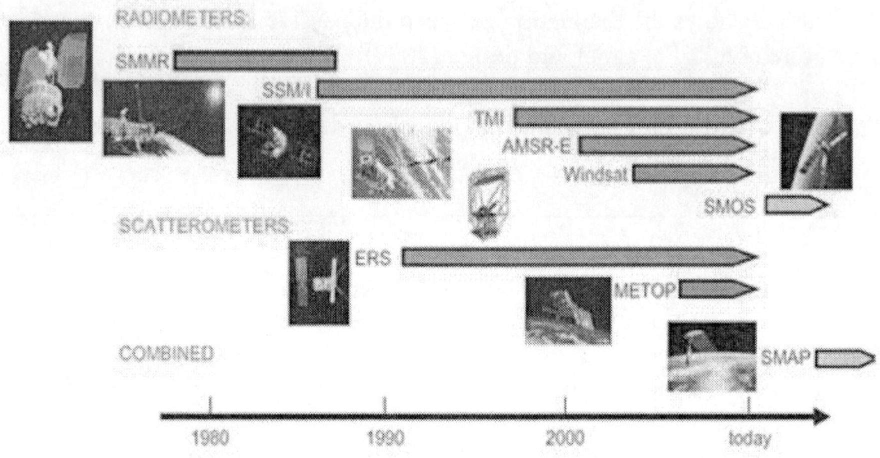

Figure 9. The increasing satellite missions that utilise microwave sensors. (Source Wagner and de Jeu, 2009).

Scatterometer

- Low spatial resolution
- Initially designed for ocean observation
- Global coverage in a few days
- Long-time operation rather recent, as compared to radiometers.

METOP

METOP is a polar orbiting satellite that measure has a suit of 13 Instruments for measuring ocean, land and atmospheric variables. ASCAT and GOME-2 are continuity instruments from ERS series launched 19 October 2006. METOP is operated by EUMETSAT as a weather and climate satellite. EUMETSAT is first polar orbiting satellite and 2nd largest after Envisat. MetOp and the US NOAA satellite both fly in Sunsynchronous, low-Earth, polar orbits. MetOp passes over the Equator (descending node) at 09.30 local time each orbit.

NOAA satellite passes over the Equator (ascending node) at 14.30 local time each orbit. Together, these two orbits maximise the coverage over which the observations are made and data is fed into weather forecasting systems.An example of the use of METOP for meteorological applications is found in the research of Gieske et al (2008). In this research, METOP data was combined with in-situ and remote sensing data to estimate Ethiopia's Lake Tana evaporation regimes.

Passive Microwave Methods: Microwave Radiometer Methods

The Passive microwave theory consists of

- Surface roughness effects on the apparent emissivity
- Vegetation effects on the apparent emissivity

The Passive microwave soil moisture retrieval algorithms are listed below:

i. Jackson et al (1999)
ii. Owe (2001)
iii. Bindlish et al (2003)
iv. Wen (2003)

Rainfall – Runoff Processes

Introduction

In Sub-Saharan Africa, the availability of accurate information on runoff is scarce. Estimation of runoff response from ungauged catchments has been an important subject of research for planning, development and operation of various water resources projects (Adib et al., 2010). This calls for good runoff models that account for spatially variable parameters such as rainfall; soil types and land use/land cover (Kumar et al., 1997).

Advances in computational power and the growing availability of spatial data from remote sensing techniques have made it possible to use hydrological models in spatial domain with GIS. Remote sensing data provides now a valuable source to overcome the traditional point scale parameterization into a much more broad spatial and temporal distribution with a promising potential for application in distributed hydrological models (Andersen, 2008).

The improvement of the quantity, quality and the resolution of remote sensing data have progressed over the last decades and hereby the perspectives of integrate remote sensing data into hydrological studies have also increased. Nowadays, remote sensing has been used in hydrological models to improve the spatial and temporal distribution of the output results (e.g. evapo-transpiration and recharge). The use of remote sensing helps in quantifying inputs into distributed hydrological models to improve the representation of the observed heterogeneity of the landscapes.

Spatially-distributed hydrological models have been increasingly applied to account for spatial variability of the main forcing variables (e.g. precipitation), physiographic characteristics of a catchment (e.g. topography, soil, land use) and detailed process calculation within a catchment (Das et al., 2008; Liang et al., 1994). Distributed models are utilized to undertake impact assessment studies (e.g. land use change and climate change studies), investigation of the influence of the spatial variability of catchment physiographic-climatic characteristics (Das et al., 2008; Liang et al., 1994), estimation of internal fluxes and state variables at high spatial resolution, and prediction in interior locations of a catchment.

Apart from physically based models, in most cases the Soil Conservation Service Curve number (SCS CN method) (SCS; 1972) also known as hydrologic soil group method is used (Moglen, 2000). This method is a versatile and popular approach for quick runoff estimation and is relatively easy to use with minimum data and it gives adequate results (Gupta and Panigrahy, 2008). Generally the model is well suited for small watersheds of less than 250km^2 and it requires details of soil characteristics land use and vegetation condition.

Case Study 4: Application of Topmodel to Simulate Streamflow for the Upper Save River Catchment of Zimbabwe Using Remote Sensing Data

The TOPographic driven MODEL (TOPMODEL) developed by Beven and Kirkby (1979) is among the different models that can be used to predict streamflow in data poor areas.The model uses topographic derivatives such as slope and topographic position that serve as important hydrological drivers (Kirkby and Chorley, 1967; Gallant and Wilson, 1996; Schmidt and Persson, 2003). Different hydrological processes such as infiltration which influence soil moisture content and runoff generation (Dunne, 1978; Wolock and Price, 1994) are affected by variations in topography. Accurate mapping of topography is therefore crucial for any topographic based hydrological modeling (Niu et al., 2005). Nowadays the adoption of improved methods of topographic characterisation is a critical for the successful application of TOPMODEL (Beven, 2001). The study by Gumindoga et al (2011b) investigated whether an ASTER (30m) DEM derived from satellite imagery can be used to derive key model parameters (e.g. Topographic Index) for simulating streamflow using TOPMODEL in the Upper Save River catchment of Zimbabwe.

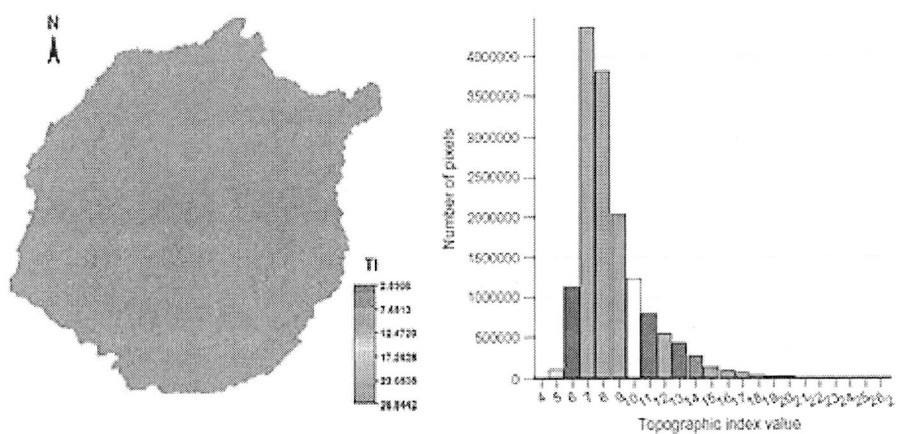

Figure 10. The Topographic Index map and frequency distribution of the topographic index value. Source Gumindoga et al (2011b).

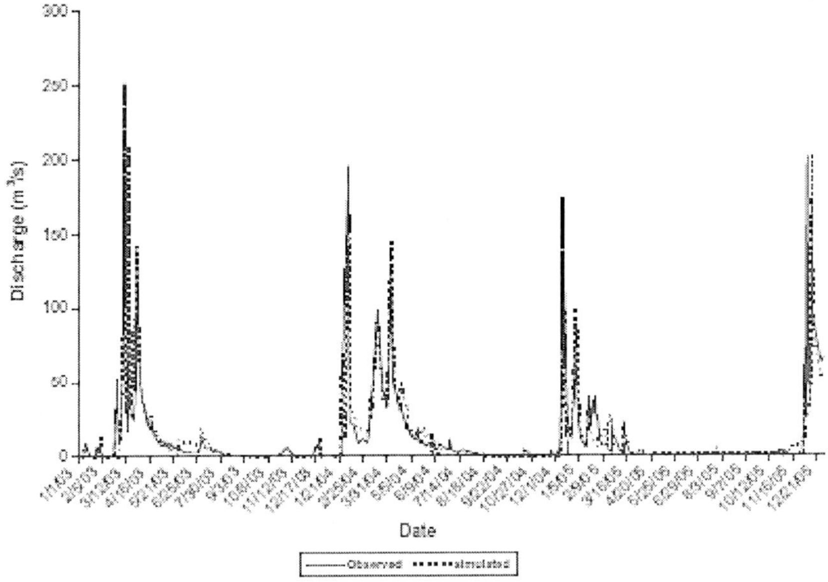

Figure 11. Calibration results for the Upper Save River catchment.

Figure 10 illustrates the topographic index of Upper Save basin and a histogram of the distribution of topographic index values. The results of the calibration process in the 2003–2005 period gave a Nash-Sutcliffe model efficiency (NS) of 0.77 and a Relative Volume Error (RVE) of 6.2% (see Figure 11). A satisfactory model performance (NS = 0.73, RVE = -8.6%) was obtained when the model was validated using 2006–2007 hydrometeorological dataset. Therefore remote sensing products such as ASTER DEM can be used to estimate model parameters for simulating streamflow.

Satellite Based Rainwater Harvesting Potential for Drought-Prone Areas

Introduction

Food insecurity in Sub-Saharan Africa is prevalent among subsistence farmers that depend heavily on rainfed agriculture (Senay and Verdin, 2004). The increased rainfall variability and drought occurrence attributed to climate change have further complicated dimensions of food security. Given the increased rainfall variability and drought prevalence in the region, there is need for increased harvesting of water resources (Gao *et al.*, 2001; Xiaoyan *et al.*, 2002).

Such water management methods and techniques may significantly improve water use efficiency and enhance sustained crop yield. Innovative GIS and remote sensing based water harvesting methods that integrates indigenous and state-of-the-art technology incorporating small-capacity ponds are important with the ultimate goal of poverty alleviation. Optimal RWH requires accurate data on rainfall-runoff processes and topographic attributes (AfDB, 2007; Munyao, 2010).

Advances in computational power and the growing availability of spatial data have made it possible to accurately predict runoff (Hadda and Yadav, 2009). The possibility of rapidly combining data of different types in a Geographic Information System (GIS) has led to significant increase in its use in hydrological applications (Senay and Verdin, 2004). The curve number method (USDA, 1984), also known as the hydrologic soil cover complex method, is a versatile and widely used procedure for runoff estimation (ICRAF, 2005).

This method requires several important properties of the watershed namely soil permeability, land use and antecedent soil moisture conditions. Thus Remote sensing (RS) and GIS provides relatively cheap and rapid methods for assessing the potential of RWH in remote and data scarce areas (Munyao, 2010).Water harvesting techniques could cushion communal farmers who rely heavily on rainfed agriculture against increased rainfall variability induced by climate change (Oduor and Gadain, 2007).

Case Study 5: GIS and Remote Sensing Based Rainwater Harvesting in Gutu District of Zimbabwe

Gutu District of Masvingo province in Zimbabwe is a typical drought prone area where improved rainfall harvesting methods may significantly improve crop yields and hence food security.

Following several years of poor rainfall, crop yields have dropped by 54 % leaving around 150,000 people in the district dependent on food aid as of 2010 (Oxfam GB, 2009). Thus optimizing rainfed crop production through water harvesting and efficient use of water resources has the potential to improve the adaptive capacity of communal farmers under increased rainfall variability regimes.

Gumindoga et al. (2011a) conducted a water balance of the district by estimating runoff using the SCS Curve number modelling technique for every 30m pixel area with the following inputs: landcover, soil, TAMSAT satellite rainfall, station rainfall and evapotranspiration (ET) data acquired and processed through GIS and remote sensing. Economic-sized ponds were determined by matching the socio-economic demands (crop production to feed a Gutu family in the off-rain season) with runoff availability, evaporation, seepages, soil type, and population density.

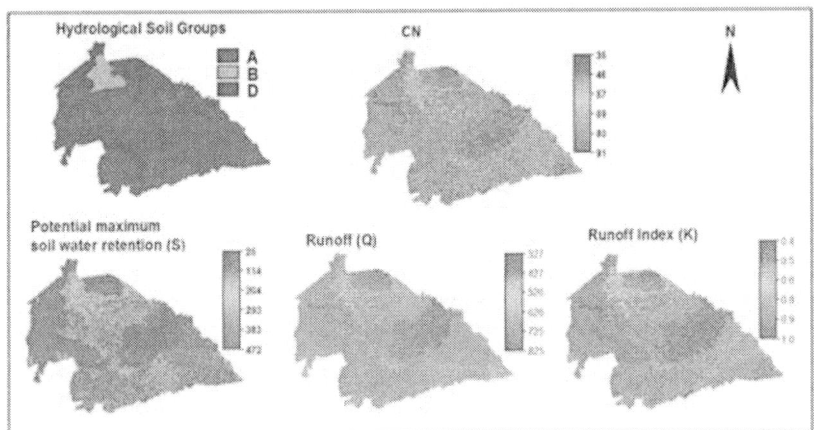

Figure 12. Runoff estimation maps.

Suitable areas for RWH were derived through integration of soil texture, Runoff Index, distance from buildings and roads, and the optimum economic sized ponds. Also, the DEM hydro processing maps for Gutu District obtained from ASTER 30m DEM and processed in ILWIS GIS software. These maps are a useful indicator of the areas suitable for RWH. Satellite-based rainfall estimates provided by the Tropical Applications of Meteorology using SATellite data (TAMSAT) for the years 2010-2011 were used for determining suitable sites for locating rain harvest ponds. There were also two important steps in the overall process – the GIS runoff estimation and determination of suitable pond dimensions for rainwater harvesting.

Figure 12 shows the the Hydrological Soil Groups, the CN map, the Potential Maximum Soil Retention, Runoff and the Runoff Index map.

Designing of Ponds for Water Harvesting

The recommended pond depth (D) for Gutu District is illustrated in Figure 13. To yield a minimum depth of 1m net supply of pond water in this district, one would need a maximum recommended design depth of approximately 2.14m. Thus 1.14m is lost to seepage and evaporation. Senay and Verdin (2004) have noted that it is the arid and semi-arid areas that require ponds of much great depth.

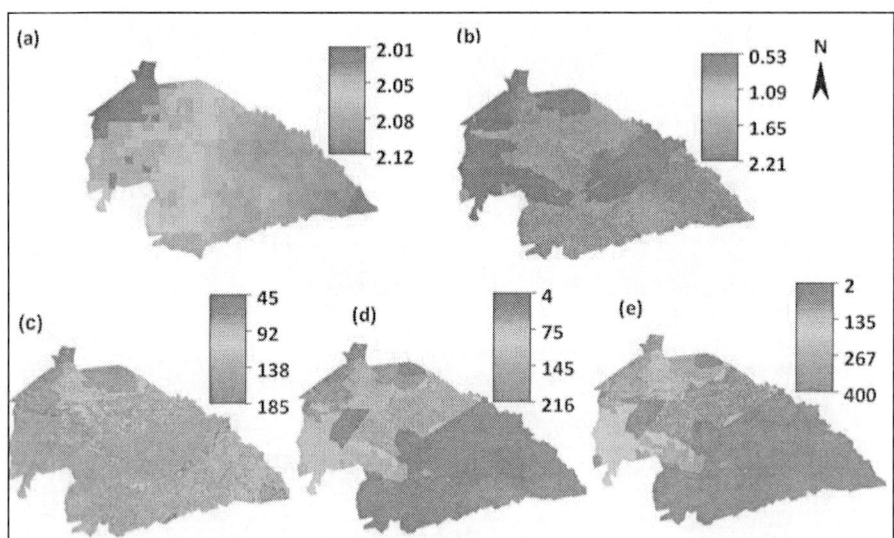

Figure 13. Spatial variation in pond depth (a), Watershed area (b), number of ponds per square kilometre (c), number of ponds per family (d) number of ponds per family normalised by watershed area (e).

runoff water for the 1000m^3 ponds and also corresponds with RWH values for Africa by Senay and Verdin (2004) and for Somalia by (Oduor and Gadain, 2007). From the study it was established that the number of ponds in the district ranged from a minium of 45to a maximum of 185 ponds per km^2. In addition, it was found that ponds per family ranged between 4 to 216 ponds per family. When the number of ponds normalised by watershed area increased to a maximum of 400 ponds.

From this case study, it can be concluded that GIS and RS can be used to acquire inputs for suitable design of ponds through the integration of Runoff availability,evaporation, seepages, soil type, population density and Gutu family food requirements.

Water Quality Assessment

Introduction

This section focuses attention on how quantitative information about the ocean, inland and coastal waters can be gained by instruments flying on satellites.

The obtained information from remote sensing sensors above an aquatic body can be schematically presented (Figure 14). Figure 15 also shows the processes that affect reflectance of natural water.

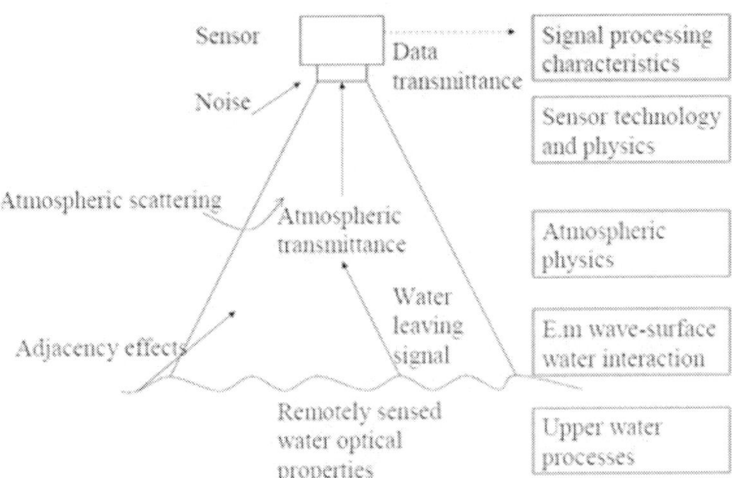

Figure 14. Information flow in remote sensing.

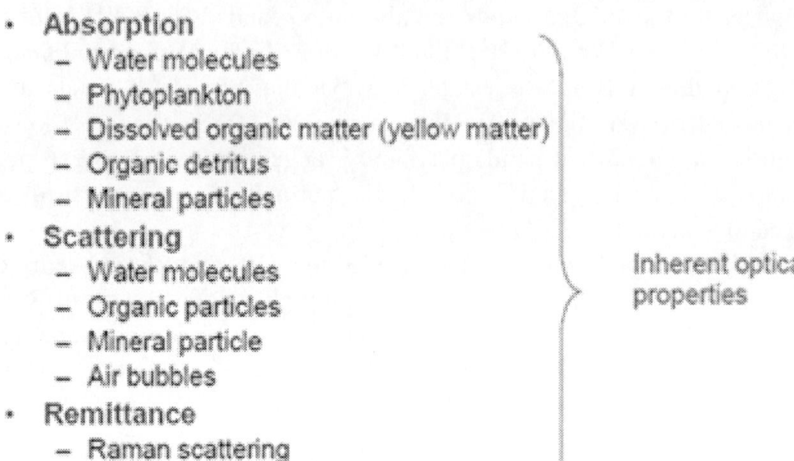

Figure 15. Processes that affect reflectance of natural water.

Optically Significant Constituents

Natural water (fresh and saline) contain suspended particles (biological and physical) and dissolved organic matter. Dissolved organic matter referred to as colored dissolved organic matter CDOM or yellow substance is created from the decaying process of organic particles. Suspended particles are of biological and physical origins. Biological particles includes: viruses, phytoplankton, zooplankton, bacteria detritus, collides and large particles (conglomerate). Physical particles are from weathering and river erosion of terrestrial rocks and soils and include: clay minerals, quartz sand and metal oxide.

Water Types And Classification (IOCCG, 2000)

i. Case 1 waters are optically governed by phytoplankton and related particles and products
ii. Case 2 waters are influenced, beside phytoplankton and related particles, by inorganic suspended particles and dissolved organic matter which are independent of phytoplankton

Optical Properties of the Water

The bulk, or large-scale, optical properties of water are conveniently divided into two mutually exclusive classes: inherent and apparent.

Inherent optical properties (IOP's) are those properties that depend only upon the medium, and therefore are independent of the ambient light field within the medium. The two fundamental IOP's are the absorption coefficient and the volume scattering function. Other IOP's include the index of refraction, the beam attenuation coefficient and the single-scattering albedo.

Apparent optical properties (AOP's) are those properties that depend both on the medium (the IOP's) and on the geometric (directional) structure of the ambient light field, and that display enough regular features and stability to be useful descriptors of the water body. Commonly used AOP's are the irradiance reflectance, the average cosines, and the various diffuse attenuation coefficients.

The use of remote sensing for water quality assessment especially for in-land water bodies in Sub-Saharan is still green area. However, there are hopes in the future works on the use of remote sensing will be undertaken in near decade. However examples of the use of remote sensing for water quality assessment of in-land water bodies include works of (Mssanzya, 2010; Olet, 2010).

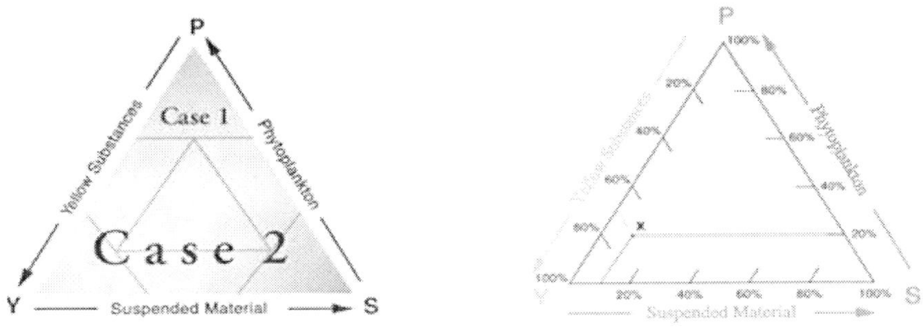

Figure 16. Case 2 waters (Adapted from IOCCG, 2000).

Case Study 6: Remote sensing of water quality in Lake Chivero, Zimbabwe

In this study Dhlamini (2012) used remote sensing to quantify the amount of Chlorophyll a, an indicator of nutrient loadings in water reservoirs, to determine the water quality levels in Lake Chivero, Zimbabwe between February-May, 2012. Chlorophyll a was determined from Moderate

Resolution Sprectro-radiometer (MODIS) derived NDVI based on empirical studies (Weghorst, 2008; Marghany and Hashim, 2010). Results from the study indicated that Chlorophyll a ranged from 0.1μg/l to 0.22 μg/l between February 2012 and April 2012 (Figure 17).

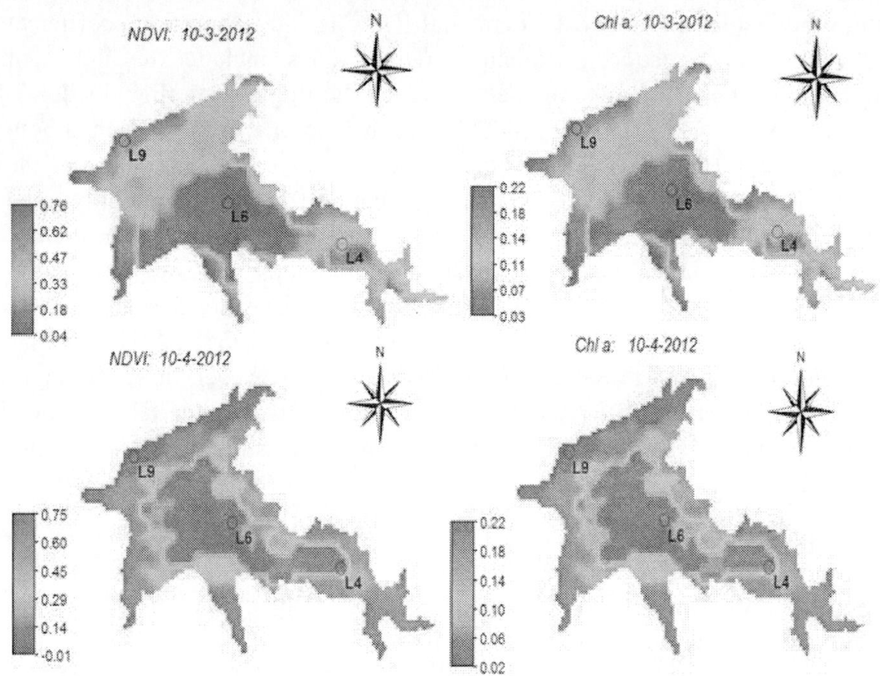

Figure 17:.Variation of NDVI and Chl a estimates in lake Chivero from March-April, 2012

Correlation analysis between satellite and in-situ measurents showed that there is a strong positive correlation ($r=0.940$, $p<0.05$) between these two measurements. Such a high correlation shows the reliability in using remote sensing to assess water quality in water bodies. A Root Mean Square Error of 0.003 further indicates that there was minimum variation in the remote sensing data and *in-situ* measurement findings, implying that satellite based measurements can be confidently be used in monitoring water quality.

Monitoring Water Levels and Flood Extents in River Basins

Introduction

The increase in the frequency and magnitude of hydrological droughts and floods requires timely and reliable data. More often ground based hydrological stations are sparsely distributed resulting in lack or insufficient data for flood prediction and drought monitoring. With recent advances in remote sensing systems such as such as Synthetic Aperture Radar (SAR), radar altimetry, ALOS, MODIS and Landsat missions, monitoring catchments from space has been greatly improved, and these data can be used to complement data from *in situ* gauging networks (Halla et al., 2010). For example, water levels and other related features such as discharge, floodplain-river connectivity, and flood extent are now being monitored using space-borne sensors. The addition of new and future satellites, such as ICESat-1 and -2, GRACE and SWOT, will guarantee the continuation of research in this field and the continued advancement of knowledge and understanding river basins (Halla et al., 2010).

One of the many fields that these technologies can be applied is to validate flood inundation models. For a long time flood extent from flood inundation models were validated using the ground truth surveys which was not very much reliable. Succesfull monitoring of river levels and mapping of flood hazard , risk and vulnerability is crucial for water resources management especially in Sub-saharan Africa where people are least prepared to deal with adverse impacts of these hazards Thus satellite data offers an opportunity to effectively deal with the aforemention environmental problems.

Case Study 7: The Spatial Variation of Flood Hazard in the Chobe-Caprivi Wetland System of the Upper Zambezi Basin

Gumindoga et al (2008) determined the spatial variation of flood hazard in the Chobe Caprivi wetland system (CCWS) of the Upper Zambezi Basin by testing whether flood hazard in the CCWS is a function of environmental factors (i.e. distance away from stream networks and height above channel). To get a clear understanding of the dynamics of flooding in the study area, the changes in inundation extent withinthe Caprivi area as well as the variations in water levels in the Zambezi River at Katima Mulilo were analysed. MODIS bands 1 and 2 images were used for mapping the area under water in the wetland between the 5th of February 2004 (before the flood) and the 31st of May 2004 (after the flood).

Supervised classification using maximum likelihood dichotomized the Caprivi area into water and non-water classes. Maximum likelihood classification was chosen because it assumes that the statistics for each class in each band are normally distributed and calculates the probability that a given pixel belongs to a specific class. Each pixel is assigned to the class that has the highest probability (that is, the maximum likelihood). If the highest probability is smaller than a threshold specified, the pixel remains unclassified (Richards, 1999).

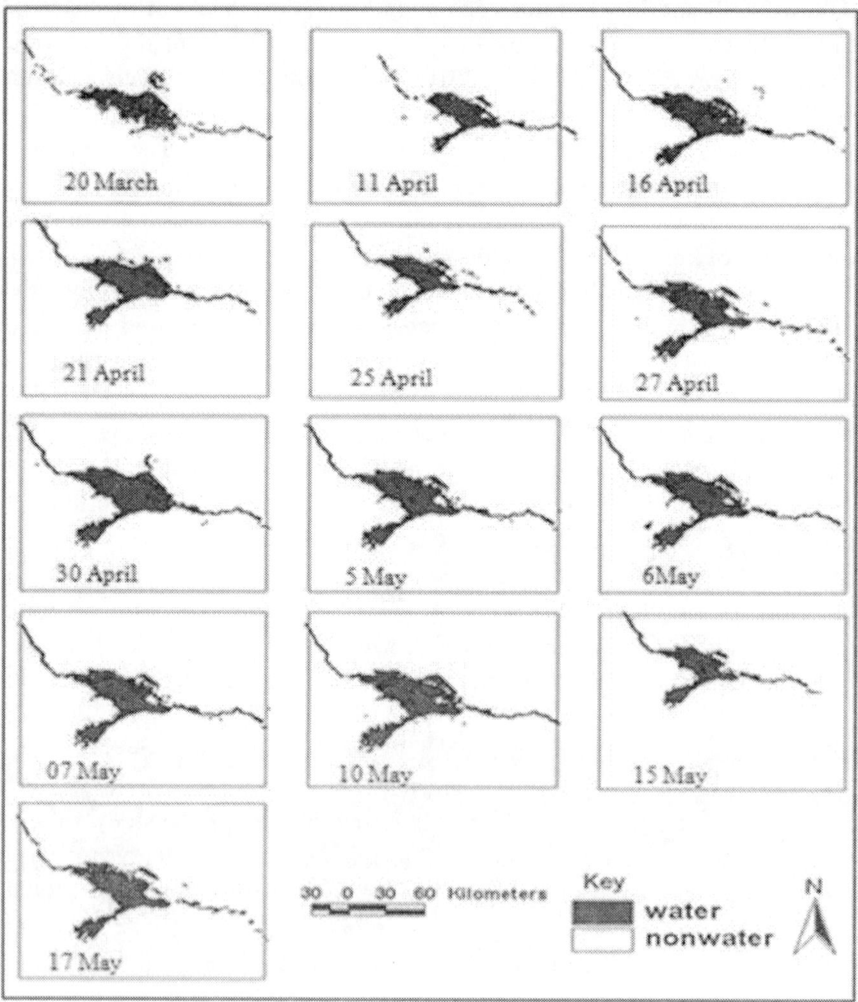

Figure 18. Changes in the area under water from 20 March to 17 May 2004 as analyzed from MODIS imagery.

Figure 18 shows changes in the area under water in the CCWS between 20 March and 31 May of the year 2004. It was observed that the least inundated areas were on 20 March whilst the highest was between the 21st and the 30th April of which the area gradually decreased until 7 May. This suggests that the Upper Zambezi River between Victoria Falls and Katima Mulilo is a constantly changing environment with half of the year experiencing rising waters while the other half has receding waters.

Figure 19 illustrates the same variations in the Caprivi Wetland water area in the form of a graph that are associated with flooding conditions. The inundated extent remained fairly the same from 10 to 17 May. Generally the trend depicts a fluctuation in the size of flooded area over the studied time.

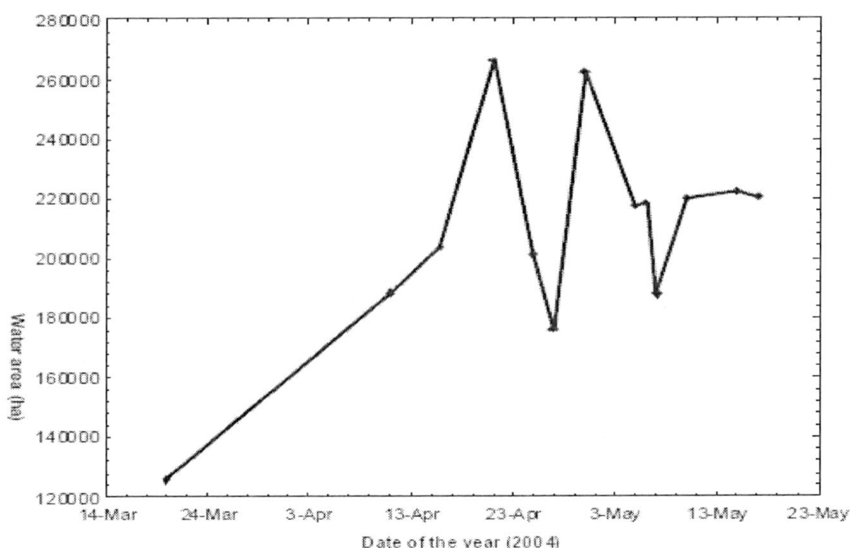

Figure 19. Inundation extent measured from MODIS imagery.

The ability of remote sensing to track changes in water extent is critical for predicting and managing floods. Such remte sensing capabilities can reduce loss of lives and property in regions such as Sub-Saharan Africa.

CONCLUSION

Remote sensing is a fast growing technology around the world. The recent real-time capacity of receiving remote sensing data proves vital for enhanced and effective water resources management. Sub-Saharan Africa is a region plagued by scarce in-situ hydrology and meteorological data and this limits professionals' ability for better planning and management of water resources. Real-time remote sensing data therefore provides an answer to fill in the gap for the scarce and missing data to improve water resources management in sub-Saharan Africa.

The chapter presents some examples on how remote sensing data has been successfully used to this extend. As observed, remote sensing data has been used exhaustively in the derivation of watersheds in many parts of Africa. Also, satellite products such as MPE, TRMM and TAMSAT are also widely available as rainfall products. Disasters such as floods have also been monitored using remote sensing data. An area which however has not been properly researched into is water quality assessment using remote sensing of in land waters in the region. This is partly due to world-wide slow pace of this area. However, the potential to use remote sensing for water quality assessment in the region proves to become vital in the coming decades especially the fact that the region has some the large river basins in the world.

In general, remote sensing proves a vital tool for effective water resources management in the region and further can aid decision makers in irrigation, flood management and achieving IWRM principles in river basin and safeguard countries against water related conflicts.

REFERENCES

Adib, A., Salarijazi, M. and Najafpour, K., (2010). Evaluation of Synthetic Outlet Runoff Assessment Models. J. *Appl. Sci. Environ. Manage,* 14 (3): 13-18.

AfDB, (2007). African Development Bank. Assessment of best practices and experiences in rainwater harvesting. *Rainwater Harvesting Handbook.* AfDB, Tunis, Tunisia.

Anornu, G.K., Kabo-bah, A.T. and Kortatsi, K., Unpublished. Comparability studies of high and low resolution DEMs for Watershed Delineation in the Tropics - *A case of the Densu River Basin of Ghana* (West Africa).

Kwame Nkrumah University of Science and Technology, Kumasi, Ghana

Bastiaanssen, W.G.M. (1998) Remote sensing in water resources management: the state of the art International Water Management Institute, Colombo, Sri Lanka.

Beven, K.J., Kirkby, M.J., 1979. A physically based, variable contributing area model of basin hydrology. Hydrol. Sci. – Bull. – des Sci. Hydrol. 24 (1),43–69.

Beven, K. and Kirkby, M.J., (1993). Channel network hydrology. John Wiley and Sons, Chichester, England. ISBN 0-471-93534-4.

Beven, K.J., 2001. Rainfall-Runoff Modelling, The Primer. John Wiley & Sons, Lancaster, UK.

Bindlish, R. et al., 2003. Soil moisture estimates from TRMM Microwave Imager observations over the Southern United States. *Remote Sensing of Environment,* 85(4): 507-515.

Blackmore, T., Chadwick, R., Francis, P., Saunders, R. and Grimes, D., (2009). Rainfall estimation over Africa using MSG.

Brutsaert, W., (2005). Hydrology: an introduction. Cambridge University Press.

Buiten, H.J. and Clevers, J.G.P.W., (1993). *Land observation by remote sensing: theory and applications.* Gordon and Breach Science Publishers.

Chen, C., Tang, S., Pan, Z., Zhan, H., Larson, M., & Jo¨nsson., L. (2007) Remotely sensed assessment of water quality levels in the Pearl River Estuary, China. *Marine Pollution Bulletin*, **54** 1267-1272.

Das, T., Bárdossy, A., Zehe, E. and He, Y., (2008). Comparison of conceptual model performance using different representations of spatial variability. *Journal of Hydrology*, 356(1-2): 106-118.

Dhlamini,.S (Unpublished) Evaluation of water quality status in Lake Chivero, Zimbabwe University of Zimbabwe, Harare, Zimbabwe

Dunne, T., 1978. Field studies of hillslope flow processes. In: Kikby, M.J. (Ed.), Hillslope Hydrol.. Wiley, London, pp. 227–294.

Gallant, J.C., Wilson, J.P., 1996. TAPES-G: a grid-based terrain analysis program for the environmental sciences. *Comput. Geosci.* 22 (7), 713–722.

Gao, S.M., Yang, F.K. and Zhang, D.W., (2001). Approaches of water resources efficient use in the gully and valley region of Loess Plateau, *Proceedings of the China National Symposium and International Workshop on Rainwater Utilization*, pp. 258-267.

Garbrecht, J. and Martz, L.W., (1999). Digital elevation model issues in water resources modeling. In: *Proceedings from invited water resources sessions,* ESRI international user conference, pp. 1-17.

Gieske, A., Rientjes, T., Haile, A.T., Abeyou, W. and Getachew, H., (2008). Determination of Lake Tana evaporation by the combined use of SEVIRI, *AVHRR and IASI,* pp. 7.

.Gumindoga, W., (2008). The spatial variation of flood hazard in the Chobe-Caprivi Wetland system of the Upper Zambezi Basin, *MSc Thesis,* University of Zimbabwe, Harare.

Gumindoga, W., Kabo-bah, A., Madamombe, C.E., Hoko, Z. and M.D, S., (2011a). Rainwater Harvesting for Improved Agriculture in Gutu District of Zimbabwe, *12th WaterNet/WARFSA/GWP-SA Symposium*, October 2011, Maputo, Mozambique.

Gumindoga, W., Rwasoka, D.T. and Murwira, A., (2011b). Simulation of streamflow using TOPMODEL in the Upper Save River catchment of Zimbabwe. *Physics and Chemistry of the Earth* 36 (2011) 806-813.

Gupta, R.K. and Panigrahy, S., (2008). Predicting the spatio-temporal variation of runoff generation in India using remotely sensed input and Soil Conservation Service Curve Number model. *Journal of Current Science,* 95: 1580-1587.

Hadda, M.S. and Yadav, R.P.A., (2009). Impacts of Small Rain Water Harvesting Tanks on Agriculture and Livelihood of Framers: A case study from Submontane Region. *Journal of Soil and Water Conservation*, 8(1): 33-66.

Halla, A.C., Schumanna, G.J.-P., Bambera, J.L. and Batesa, P.D., (2010). Tracking water level changes of the Amazon Basin with spaceborne remote sensing and integration with large scale hydrodynamic modelling: a review. *Physics and Chemistry of the Earth,* doi: 10.1016/j.pce.2010. 12.010.

Hong, Y., Hsu, K.-L., Sorooshian, S., & Gao, X. (2004) Precipitation Estimation from Remotely Sensed Imagery Using an Artificial Neural Network Cloud Classification System. *Journal of Applied Meteorology*, **43**, 1834-1851.

Huggett, R.J., (2003). *Fundamentals of Geomorphology*. Routledge, London, UK.

ICRAF, (2005). Potential for Rainwater Harvesting in Africa. *A GIS overview,* ICRAF and UNEP.

Inglada, J. & Giros, A. (2004) On the real capabilities of remote sensing for disaster management-feedback from real cases In Geoscience and Remote Sensing Symposium, Vol. 2, pp. 2. IEEE, Piscataway, NJ.

IOCCG, (2000). Remote Sensing of Ocean Colour in Coastal, and Other Optically-Complex, Waters. Sathyendranath, S. (ed.), *Reports of the International Ocean-Colour Coordinating Group,* No. 3, IOCCG, Dartmouth, Canada.

Johnson, G.E., Achutuni, V.R., Thiruvengadachari, S., & Kogan, F., eds. (1993) *The role of NOAA satellite data in drought early warning and monitoring. Drought Assessment, Management, and Planning: Theory and Case Studies*. Kluwer Academic.

Kabo-bah, A.T., Madamombe, C.E. and Rwasoka, D.T., (2011). Estimation of hyper-temporal evapotranspiration over the middle-zambezi using the GEONETCAST toolbox and SEBS., *12th WaterNet/WARFSA/GWP-SA Symposium,* 26-28 October 2011, Maputo, Mozambique.

Kabo-bah, A.T., Yuebo, X., Andoh, R. and Odai, S.N., (2012). A Comparison of TAMSAT Satellite Rain Derived Products with Meteorologically Measured Rainfall in the Volta Basin (West Africa), *7th International Conference on Water Sensitive Urban Design,* 21–23 February 2012, Melbourne, Australia.

Kenward, T., Lettenmaier, D.P., Wood, E.F. and Fielding, E., (2000). Effects of Digital Elevation Model Accuracy on Hydrologic Predictions. *Remote Sensing of Environment,* 74(3): 432-444.

Kim, Y.J., Evans, R.G., & Iversen, W.M. (2008) Remote Sensing and Control ofan Irrigation System Using a Distributed Wireless Sensor Network. *IEEE Transactions on Instrumentation and Measurement*, **57**, 1379-1387.

Kirkby, M.J., Chorley, R.J., 1967. Throughfow, overland flow and erosion. Bull. *Int.Assoc. Sci. Hydrol.* 12, 5–21.

Kumar, P., Tiwari, K.N. and Pal, D.K., (1997). Establishing SCS runoff curve number from IRS digital database. *Journal of Indian society of Remote sensing,* 19: 246-251.

Kusangaya, S., (2007). *Water Recourses Management lecture Notes,* Dept of Geography and Environmental Scince UZ, Harare, Zimbabwe.

Liang, X., Lettenmaier, D.P., Wood, E. and Burges, S., (1994). A simple hydrological based model of land surface water and energy fluxes for general circulation models *Journal of Geophysical Research*, 99(D7): 14 415-14 428.

Lillessand, T.M. and Kiefer, R.W., (1994). *Remote Sensing and Image Interpretation,* 3rd (Eds) (New York: John Wiley and Sons).

Marghany, M. and Hashim, M., 2010. MODIS satellite data for modeling chlorophyll *a* concentrations in Malaysian coastal waters. International Journal of the Physical Sciences, Vol. 5(10): pp 1489-1495.

Moglen, G.E., (2000). Effect of orientation of spatially distributed curve number in runoff calculations. *Journal of American Water resource Association,* 36: 1391-1400.

Mssanzya, S.M., (2010). Monitoring and predicting eutrophication of inland waters using remote sensing*., University of Twente Faculty of Geo-Information and Earth Observation ITC, Enschede.*

Munyao, J.N., (2010). Use of Satellite Products To Assess Water Harvesting Potential In Remote Areas Of Africa. A Case Study of Unguja Island, Zanzibar Enschede, 80 pp.

Niu, G., Yang, Z., Dickson, R.E., Gulden, L.E., 2005. A simple TOPMODEL-based runoff parameterization (SIMTOP) for use in global climate models. J. Geophys. Res. 110, D21106

Oduor, A.R. and Gadain, H.M., (2007). Potential of Rainwater Harvesting in Somalia, A Planning, Design, Implementation and Monitoring Framework, Technical Report NoW-09, 2007,FAO-SWALIM, Nairobi, Kenya.

Olet, E., (2010). Water Quality Monitoring of Roxo reservoir using Landsat Images and In-situ Measurements.

Owe, M., de Jeu, R. and Walker, J., (2001). A methodology for surface soil moisture and vegetation optical depth retrieval using the microwave polarization difference index. Geoscience and Remote Sensing, *IEEE Transactions on*, 39(8): 1643-1654.

Oxfam GB, (2009). EIA Report for the Proposed Irrigation Project at Ruti Dam, Gutu. *Report prepared by Institute of Mining Research and Department of Geography and Environmental Science,* University of Zimbabwe.

Ozesmi, S.L. & Bauer, M.E. (2002) Satellite remote sensing of wetlands. *Wetlands Ecology and Management*, **10**, 381-402.

Quinn, P.F., Beven, K.J. and Lamb, R., (1995). The Ln (a/TanB) Index: How to calculate it and how to use it within the Topmodel framework. *Hydrological Processes*, 9: 161-182.

Rees, W.G., 2006. Remote Sensing of Snow and Ice, Taylor and Francis, 285 pages.

Ritchie, J.C. and C.M. Cooper, 1998. Comparison of Measured Suspended Sediment concentration with Suspended Sediment Concentrations Estimated Landsat MSS data. International Journal of Remote Sensing 9(3): 379-387.

Sabins, F.F., (2007). *Remote sensing: principles and interpretation.* Waveland Press.

Saraf, A.K. & Choudhury, P.R. (1998) Integrated GIS and Remote Sensing for groundwater exploration and Identification of artificial ground water recharge sites. *International Journal of Remote Sensing*, **19**, 1825-1841.

Schmidt, F., Persson, A., 2003. Comparison of DEM data capture and topographic wetness indices. *Prec. Agric.* 4 (2), 179–192.

Senay, G.B. and Verdin, J.P., (2004). Developing Index Maps of Water-Harvest Potential in Africa. *Applied Engineering in Agriculture*, 20(6): 789-799.

USDA, 2004. Estimation of Direct Runoff from Storm Rainfall, National Engineering Handbook, pp. 79

Wagner, W. and de Jeu, R., (2009). *Presentation at the Kick-off of WACMOS project.*

Wagner, W., Lemoine, G. and Rott, H., (1999). A Method for Estimating Soil Moisture from ERS Scatterometer and Soil Data. *Remote Sensing of Environment,* 70(2): 191-207.

Weghorst, P., L., 2008. MODIS algorithm assessment and principal component analysis of chlorophyll concentration in Lake Erie, Kent State University, USA

Wen, J., Su, Z. and Ma, Y., (2003). Determination of land surface temperature and soil moisture from Tropical Rainfall Measuring Mission/Microwave Imager remote sensing data. *J. Geophys. Res.,* 108: 4038.

Wolock, D.M., Price, C.V., 1994. Effects of digital elevation model map scale and data resolution on a topography-based hydrologic model. Water Resour. Res. 30,3041–3052.

Xiaoyan, L., Ruiling, Z., Jiadong, G. and Zhongkui, X., (2002). Effects of Rainwater Harvesting on the Regional Development and Environmental Conservation in the Semiarid Loess Region of Northwest China, *12th ISCO Conference Beijing.*

In: Remote Sensing
Editor: Enner Alcântara
ISBN: 978-1-62417-140-6

Chapter 4

ESTIMATING PARTICLE DISPERSION AND RESERVOIR TRANSPORT PROCESSES FROM SATELLITE TRACKED DRIFTERS DATA: METHODOLOGY AND APPLICATIONS

***Arcilan Assireu**[*1]**, Felipe Siqueira Pacheco**[2,3]*
***and Camila Baldini**[1]*
[1]Natural Resources Institute –
Federal University of Itajubá –Pinheirinho, Itajubá, Brazil
[2]National Institute for Space Research - Jardim da Granja,
São José dos Campos, São Paulo, Brazil
[3]Federal University of Juiz de Fora, Juiz de Fora, Brazil

ABSTRACT

Satellite tracked drifters were deployed in order to study the horizontal transport and dispersion in the epilimnion of a large hydroelectric reservoir. Our study was motivated by the dispersion implications on the reservoir surface process, such as larval dispersion, turbulent mixing, point-source fertilizer, dispersion of pollutants, and many others. A review of methods used to estimate the dispersion and transport of surface water is presented. From the perspective of fertilization the focus is horizontal dispersion in the epilimnion. The large

[*] E-mail: arcilan@unifei.edu.br

difference between vertical and horizontal length scales in the epilimnion allows us to assume that vertical mixing occurs relatively quickly. Our results have demonstrated the effectiveness of the drifter in estimating differential kinematic properties and horizontal dispersion. A discussion of how these parameters can be useful for the safe and efficient management of the waterway is carried out. While so far no evidence exists about near-inertial motions in the lakes and reservoirs, it will be shown that this important mixing process could be occur in lakes and reservoirs where the effect of the Earth´s rotation on the water body´s response is important. We will exemplify this theory and application with a case study for Brazil Current. Near-inertial motions on the Brazilian Shelf Break (BSB) were monitored using surface drifters. Strong and superinertial currents were generated by wind burst up to 12 m s-1. The near-inertial current amplitude was up to 40 cm s-1 at the 15m depth. The observed near-inertial frequency along the BSB was about 10% higher than the local inertial frequency, suggesting that this near-inertial motion was embedded in a background negative vorticity field. The cyclonic relative vorticity (negative) was able to shift the Effective Inertial Frequency higher (blue shift), towards the diurnal frequency, creating conditions to resonance between the near-inertial currents and the daily wind cycle. This result is supported by an analytical model which indicated also that, in the scenario of background positive vorticity, there are effective conditions to removing near-inertial energy from the mixed layer. It must explain why only 1 out of 42 drifters presented an enhanced near-inertial motion. The frequency range in which the spectral density follows a -5/3 Kolmogorov scaling (i.e., the inertial subrange from turbulence) was investigated and some speculations about the interaction between near-inertial motions and mean flow is presented.

Keywords: Satellite tracked drifters, particle dispersion, horizontal transport

1. INTRODUCTION

The use of new Technologies has enabled the synoptic measurement of the current in lakes and reservoirs, in a way not seen before. The remote sensing of the sea surface properties is a routine tool to obtain information on currents, waves, turbulence, and oil spills. Knowledge of the processes related to horizontal dispersion is very important to understanding the biogeochemical process of a water body. Satellite tracked drifters trajectories are a useful tool for understanding and diagnosing the motion of the ocean, lake and reservoir. This allows us to obtain information about transport and dispersion and, consequently, about destination of a water particle besides the spread of several particles. This information is relevant for estimating the trajectories and rate of evolution of oil spills (Soomer et al., 2010), larval dispersal and other living organisms (Corell et al., 2011), and for planning rescue operations or finding lost goods.

Despite this, very few horizontal dispersion studies exist for lakes and reservoirs (Lawrence et al. 1995; Stocker and Imberger, 2003; Pacheco et al., 2011). In the context of this study the use of remote sensing techniques to evaluate a lake or reservoir current is relatively new.

Many problems of environmental contamination of rivers, lakes, reservoir, sea and atmosphere can often be simplified to the solution of a mathematical model involving both advective and diffusive elements. The advection describes the mechanism of entrainment of a pollutant in the ambient flow and its transport in solution or suspension, with velocity equal to the flow velocity. The other mechanism is turbulent motion, normally associated with random distribution of velocity. In nature the flows tends to present unsteady and heterogeneous turbulent behavior.

Heterogeneous flow field means that the velocity vary spatially while turbulent means that the instantaneous velocity varies randomly related both time and space about some mean value. The usual decomposition of turbulent motion in a mean value and a fluctuating component is a not easy task because of the uncertainty in defining a mean value of flow. Nevertheless, all numerical dispersion models are based on the parameterization of the turbulent transport of matter through the use of a non-physical quantity named eddy diffusivity coefficients (EDC). Thus, the parameterization of the turbulent transport of matter is based on this. Okubo and Ebbesmeyer (1976) (OE) have presented a method to provide a measure of the turbulent induced diffusion. Their methods use a spatial array of measurement of simultaneous currents and decompose these into their mean velocities and gradients. These quantities are subtracted from the original measurements to obtain the residual turbulent velocity field, form which the turbulent induced EDC can be extracted.

2. Material and Methods

The drifters used here were drogued at depths of 1m, because the focus of our investigation was the horizontal mixing process in the epilimnion. The sampling frequency was 0.1 Hz and the precision was 10-m horizontal root mean square.

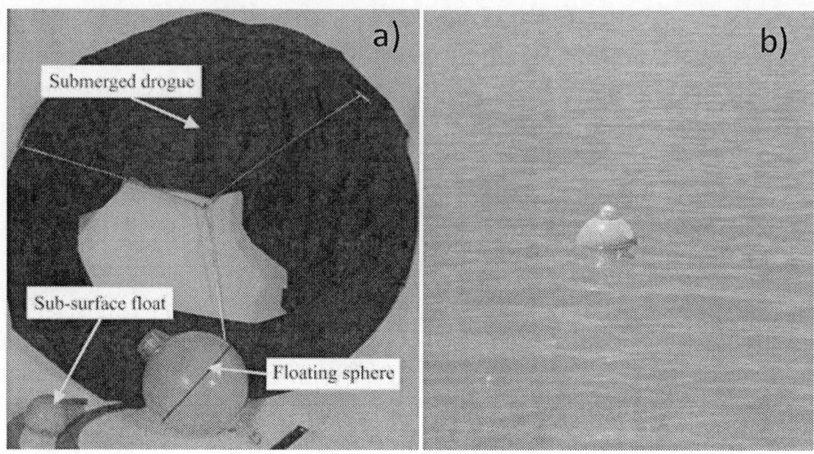

Figure 1. a) drift used in this work (disassembled) b) drift deployed in Furnas reservoir.

Discrepancies between the motion of the drifters and that of water parcels caused by factors such as wind induced slippage and the finite size of the drogues were summarized, for example, by Niiler et al. (1995). They showed that for an area ratio larger than 40, wind-induced slippage is less than 1 cm s-1 in a 10 m s-1 wind. For more detail, see Johnson et al. (2003).

The drifters used in this work are composed of two parts: a floating one and a submerged one. The floating part consists of a sphere made of fiberglass with diameter of 21 cm. Within, batteries, GPS, digital memories and electronic boards are installed. The sub-surface float is a small hollow sphere made of fiberglass which is intended to provide greater stability. The submerged part is a cylindrical structure (called drogue) in order to attach the drifter to the currents. This drogue is made of nylon and stainless steel and has several small holes to improve hydrodynamics (Figure 1). An adjustable cable connects the submerged part to the immersed part of the drifter. The ratio between the drag area of the drogue and the surface is 40 to 1 according to the literature (Niiler et al., 1995).

2.1. Differential Kinematic Properties – Okubo-Ebesemeyer´s Method

Consider positions obtained from n drifter´s observed simultaneously at time m. The resulting time series of position are function of time t with increments k can be represented in a Lagrangian frame as (Okubo et al. 1976):

$$\begin{cases} x_i = x_i\,(a_i, b_i, k),\ i = 1,2,3,\ldots n \\ y_i = y_i\,(a_i, b_i, k),\ k = 1,2,3,\ldots m \end{cases}$$

The instantaneous positions xi, yi are Taylor expanded about the centroid:

$$\begin{cases} x_i(a_i,b_i,k) = x(\bar{a},\bar{b},k) + \left(\frac{\partial x}{\partial a}\right)_0 (a_i - \bar{a}) + \left(\frac{\partial x}{\partial b}\right)_0 (b_i - \bar{b}) + x_i''(a_i,b_i,k) \\ y_i(a_i,b_i,k) = y(\bar{a},\bar{b},k) + \left(\frac{\partial y}{\partial a}\right)_0 (a_i - \bar{a}) + \left(\frac{\partial y}{\partial b}\right)_0 (b_i - \bar{b}) + y_i''(a_i,b_i,k) \end{cases} \tag{1}$$

The equivalent instantaneous surface velocity in the form of (1) can be, using matricial notation, represented as:

$$U = U_0 + (R - \bar{R})\psi + U'', \tag{2}$$

where the position matrices is

$$\mathbf{R}\text{-}\overline{\mathbf{R}} = \begin{bmatrix} x_1 - \bar{x} & y_1 - \bar{y} \\ x_2 - \bar{x} & y_2 - \bar{y} \\ \vdots & \vdots \\ x_n - \bar{x} & y_n - \bar{y} \end{bmatrix} \tag{3}$$

Instantaneous speed matrices:

$$\mathbf{U} = \begin{bmatrix} u_1 & v_1 \\ u_2 & v_2 \\ \vdots & \vdots \\ u_n & v_n \end{bmatrix} \tag{4}$$

The mean current properties:

$$\boldsymbol{\psi}(k) = \begin{bmatrix} \left(\frac{\partial \mathbf{u}}{\partial \mathbf{x}}\right)_0 & \left(\frac{\partial \mathbf{v}}{\partial \mathbf{x}}\right)_0 \\ \left(\frac{\partial \mathbf{u}}{\partial \mathbf{y}}\right)_0 & \left(\frac{\partial \mathbf{v}}{\partial \mathbf{y}}\right)_0 \end{bmatrix} \quad (5)$$

and the instantaneous turbulence matrices

$$\mathbf{U}''(\mathbf{R}, k) = \begin{bmatrix} u_1'' & v_1'' \\ u_2'' & v_2'' \\ \vdots & \vdots \\ u_n & v_n \end{bmatrix} \quad (6)$$

The Equation 2 can be re-arranged as follow:

$$\begin{cases} \psi = E^{-1}\dot{E} \\ U'' = (\dot{R}'' - \overline{\dot{R}''}) - (R'' - \overline{R''})\psi \end{cases} \quad (7)$$

where E is the lagrangian deformation rate:

$$\mathbf{E}(k) \equiv \begin{pmatrix} e_{11}(k) & e_{21}(k) \\ e_{12}(k) & e_{22}(k) \end{pmatrix}$$

, With,

$$e_{11} \equiv \left(\frac{\partial x}{\partial a}\right)_0, \quad e_{12} \equiv \left(\frac{\partial x}{\partial b}\right)_0, \quad e_{21} \equiv \left(\frac{\partial y}{\partial a}\right)_0, \quad e_{22} \equiv \left(\frac{\partial y}{\partial b}\right)_0,$$

Are the first order terms of lagrangian deformation.

From their physically more meaningful combination we have the differential kinematic properties (DKP):

$$\begin{cases} \text{Horizontal Divergence: } \gamma = \left(\dfrac{\partial \mathbf{u}}{\partial \mathbf{x}}\right)_0 + \left(\dfrac{\partial \mathbf{v}}{\partial \mathbf{y}}\right)_0 \\ \text{Vertical Relative Vorticity: } \xi = \left(\dfrac{\partial \mathbf{v}}{\partial \mathbf{x}}\right)_0 - \left(\dfrac{\partial \mathbf{u}}{\partial \mathbf{y}}\right)_0 \\ \text{Horizontal Stretching Deformation Rate: } \alpha = \left(\dfrac{\partial \mathbf{u}}{\partial \mathbf{x}}\right)_0 - \left(\dfrac{\partial \mathbf{v}}{\partial \mathbf{y}}\right)_0 \\ \text{Horizontal Shear Deformation Rate: } h = \left(\dfrac{\partial \mathbf{v}}{\partial \mathbf{x}}\right)_0 + \left(\dfrac{\partial \mathbf{u}}{\partial \mathbf{y}}\right)_0 \end{cases}$$

More details about the outline of the mathematical formulation of the Okubo-Ebesemeyer´s Method can find in Sanderson (1995), Okubo et al. (1976) among others.

Molinari and Kirwan (1975) presented a discussion about phenomenological aspects related to the DKP. Vorticity can be related to external excitation functions of circulation such as wind stress and bottom topography. Divergence, for example, is a diagnostic of vertical motion. The deformations are important in the evolution of frontal zones.

The turbulent terms in Equation 7 is composed by two terms: one related to real heterogeneity of the velocity gradients (turbulence) and other is due to measurement error in the drifter´s positions. A usual assumption is that the turbulent fluctuations and measurement errors are uncorrelated. Thus, if the turbulent fluctuation , are greater than variance of the measurement error, then we can attribute to real heterogeneity the total variation.

The standard deviation of the turbulent velocities is given by

$$\sigma_u = \sqrt{\frac{\sum_{i-1}^{n} u_i''^2}{n-3}} \quad \text{and} \quad \sigma_v = \sqrt{\frac{\sum_{i-1}^{n} v_i''^2}{n-3}} \tag{8}$$

The use of the factor n-3 instead n-1 in the equation 8 is to form the unbiased estimates of the standard deviations (Sanderson et al., 1988). This is necessary because the number of degrees of freedom associated with Equation 7 is n-3 (there are three unknowns) when there are n observations. This correction is important when the number of drifters is small (typically less than 10).

The eddy diffusivity coefficients (EDC) k_x and k_y may be obtained by analogy to a combination of mixing length scale and turbulence, interpreted as the residual velocities computed from cluster analysis (Okubo and Ebbesmeyer, 1976). Thus, the turbulence intensity was assumed to be proportional to the standard deviation of the residual velocities (σ_u, σ_v) and the mixing length scale proportional to the standard deviation of the drifter´s positions (σ_x, σ_y).Thus, for each cluster cluster $k_x = c\sigma_x\sigma_u$, $k_y = c\sigma_y\sigma_v$.

Okubo and Ebbesmeyer (1976) argued that the proportionality constant c should be of order 0.1 and 1, however, in a review of the OE´s method, Sanderson (1995) demonstrated that the more adequate order of magnitude of this constant must be 1.

The dispersion coefficients, defined for example as $k_x = 0.5\frac{d\sigma_x^2}{dt}$ (Stocker and Imberger, 2003) for the direction x, can be computed in the same form for k_{yx} and k_y . σ_x and σ_y are the standard deviation along the x and y direction, respectively, whereas $\sigma_{xy} = (\sigma_x^2 + \sigma_y^2)^{\frac{1}{2}}$.

3. Results and Discussion

3.1. Meteorological Forcing at the Study Area

Marked diel patterns were observed in air temperature, relative humidity, and wind at the study area (Figure 2). Maximum air temperature was 29°C whilst minimum air temperature was 20°C. Calm conditions normally occurred during the night and persisted up to 0900 h. In the afternoon wind speeds reach a mean value of 5 ms-1. Relative humidity was least during the afternoon and was > 60% at other times.

Except for a brief period in the late afternoon, air temperatures were generally > 2°C cooler than the surface waters. Consequently, the boundary layer above the air-water interface was unstable.

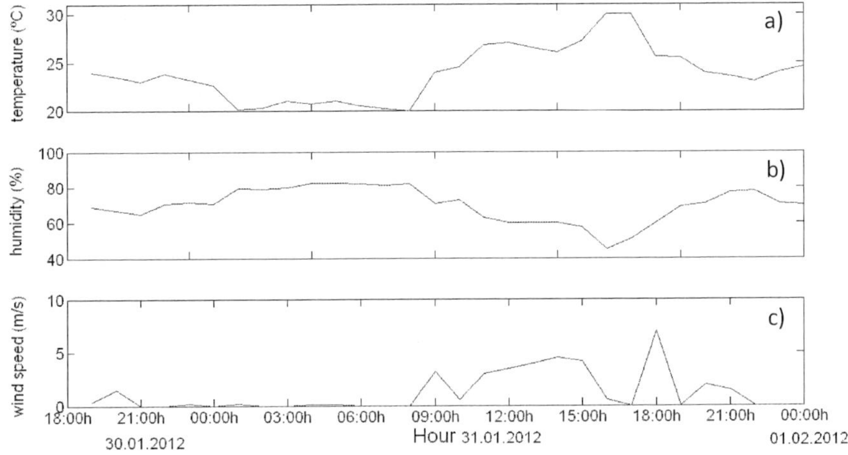

Figure 2. Air temperature (a), relative humidity (b), and wind (c) during the experiment 3 (see Table 1).

The diurnal cycle of heating and cooling and the typical stratification are illustrated in Figure 3. Strong diurnal heating, with surface temperatures reaching 29°C (Figure 4), occurred between 12 00 h and 15 00 h. During afternoon winds began and the water column began to lose heat and an increasing mixing process eroded the stratification.

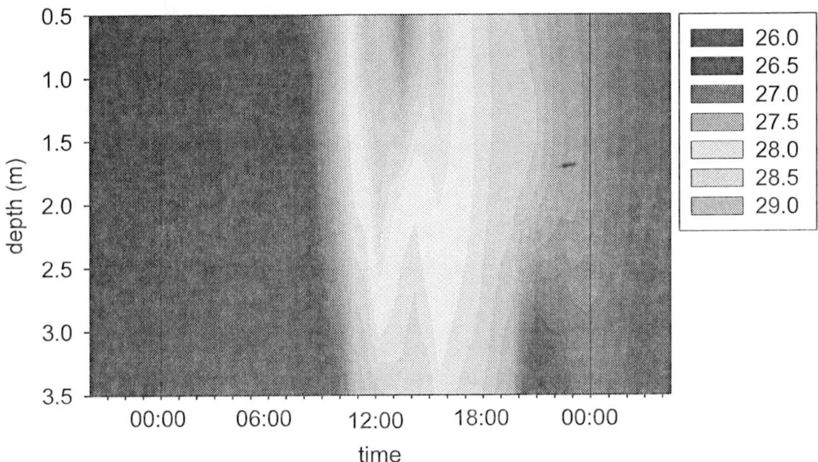

Figure 3. Water temperature profile for a margin station acquired from thermistor chain.

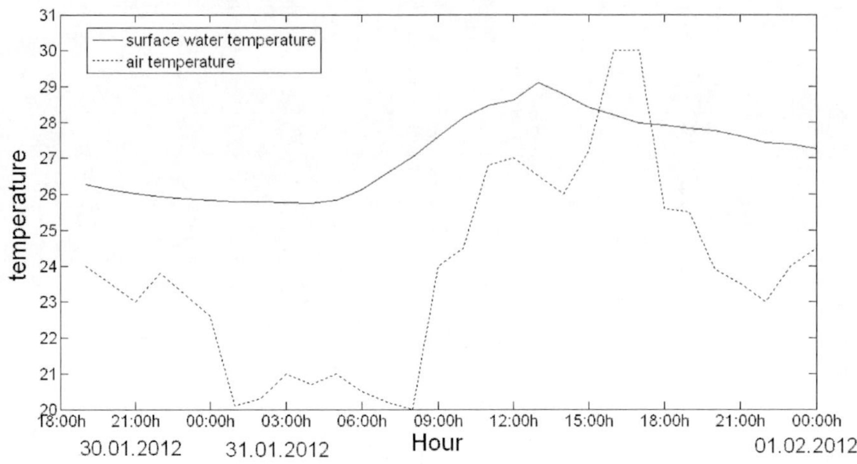

Figure 4. Air temperature and water surface temperature.

3.2. Drifters Experiment

Our drifter dispersion experiments were performed in a sector of the Furnas reservoir (21°25'S, 46°08'W), where there are projects for fish farming and the construction of a waterway (Figure 5).

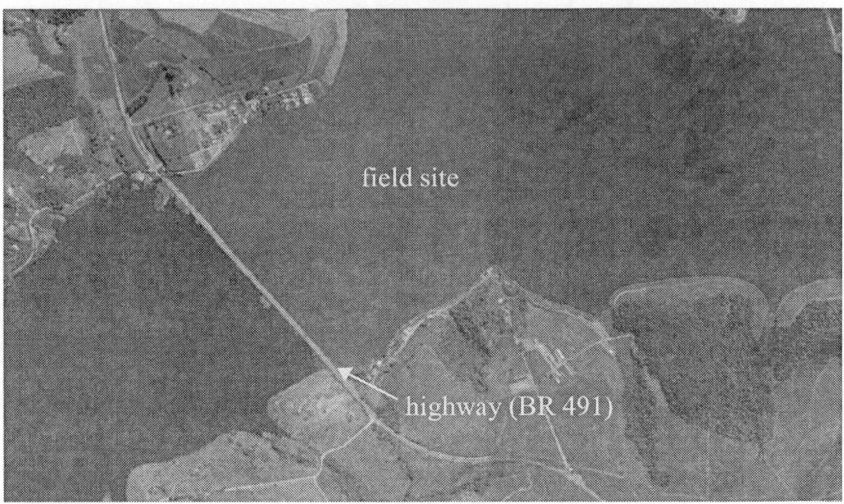

Figure 5. Field site in Furnas reservoir.

Formed in the early 60's, with the construction of the dam, the lake of Furnas is the largest body of water of Minas Gerais state, Brazil - covers an area of 1,457.48 kilometers square. It is being studied the implementation of a waterway in this lake, with its principal link the cities of Formiga and Alfenas, a distance of 250 km. It will also be built ports along the route, seeking to assist other cities in the shaft.

The release of pollution in aquatic systems may have a dramatic impact on local ecosystems (Prahl et al., 1984; Rice et al., 1993; Lekien et al., 2005), mainly if the pollution recirculates in a near close-boundary flow, as it the case in lakes and reservoirs.

In order to test the sensitive dependence of Lagrangian particle motion on initial conditions, identical releases of drifters at the same time but from two slightly different locations was made (Figure 6).

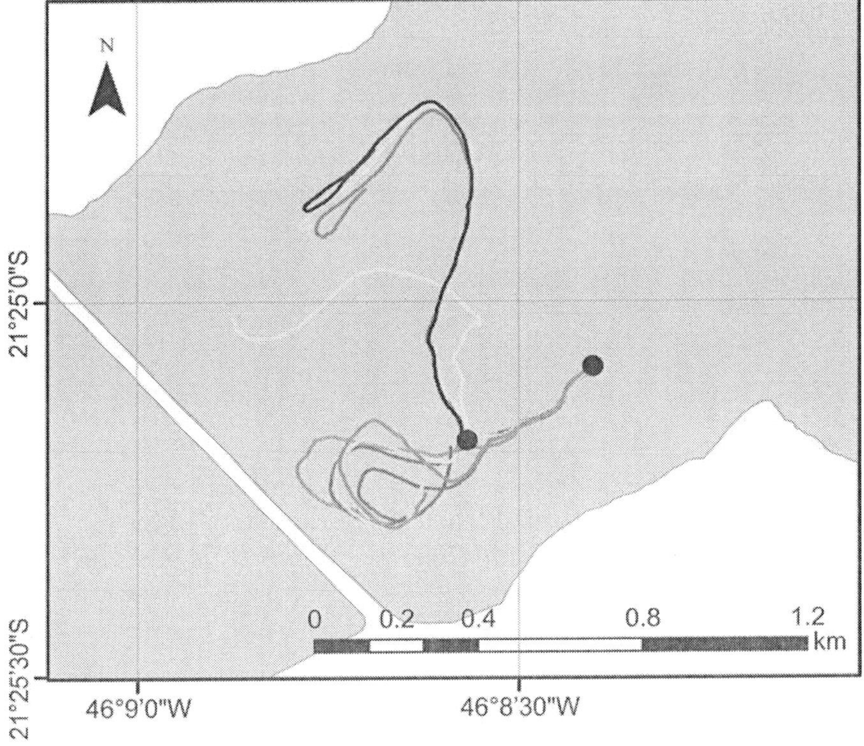

Figure 6. Trajectories of the drifters in experiment 1. Identical releases of drifters at the same time but from two slightly different locations (Circles indicate locations of deployments).

Table 1. Details of the three drifter experiments carried out in the Furnas Reservoir in 2011. Experiment were carried out at high level period (experiment 1), receding level period (experiment 2), and rising period (experiment 3). Texp is the duration of the experiment and Ttot is the total number of drifter hours available. Drifters were drogued at 1-m depth

Experiment	Drifters	Deployment	Retrieval	T_{exp} (h)	T_{tot} (h)
1	1,2,3,4,5,6,7,8,9	1900 h, 17 August	2000 h, 18 October	25	212
2	1,2,3,4,5	1930 h, 31 October	2030 h, 1 November	23	108
3	1,2,3,4,5	1930 h, 30 January	0800 h, 2 February	36	96

Table 2. Mean values of intensity of the currents

	V (cms^{-1})	σ (cms^{-1})
Campo 1 - 1	1.35	0,04
Campo 1 - 2	1.71	0,21
Campo 2	2.18	0,36
Campo 3	1.63	0,15

Concepts of sensitivity to initial conditions are related not only to location but different time of drifters release are also subject to sensitive dependence on their initial conditions. Operational water reservoirs are used to manage the supply of electricity according to the electricity demands at a certain region. Hence such reservoirs are characterized by a large variability in the water level due to variations in rain regimes and withdrawal of water. In order to evaluate the implications of this time water level variation to diffusive process, the drifters release were carried out in August 2011 (high level period), October 2011 (receding level period) and January 2012 (rising period).

Experiment 1 – Predominant direction were southwestward for the initial 10 h, showing a clear anticyclone circulation after this for drifters 1,2,3,4 and 5 (Figure 7a-1).

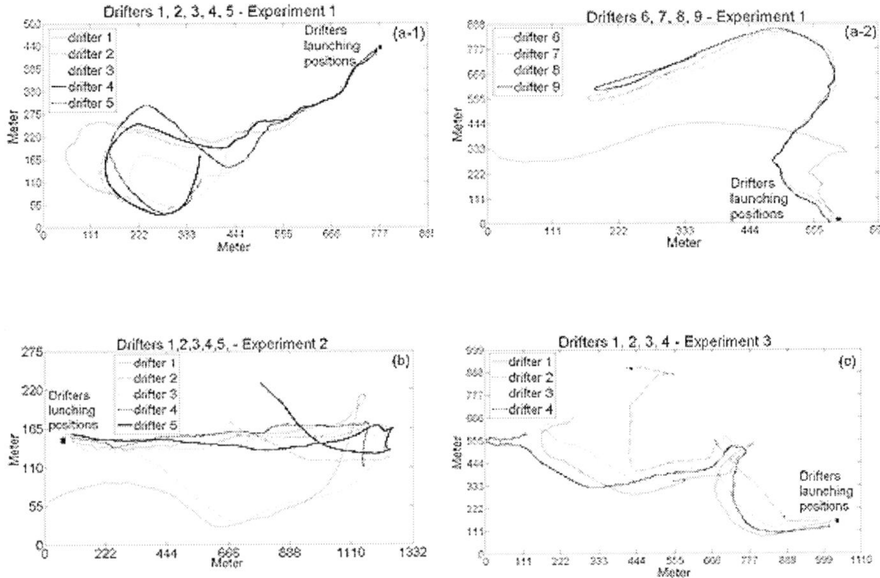

Figure 7. Trajectories of the drifters in experiment 1 (a), 2 (b) and 3 (c). Square indicate locations of deployment. The depths and dates of each experiment are given in Table 1.

Curiously, the drifter cluster released at the same time but at position slightly different, showed trajectories very different (Figure 7a-2), indicating the sensitive to initial conditions. The mean velocities were 1.3 and 1.7 cms^{-1}, respectively (Table 2).

Experiment 2 – Predominant direction were northestward for the initial 6 h, showing a clear anticyclone circulation after this for drifters 1, 2, 3, and 4 (Figure 7b). The drifter 5 presented an intricate trajectory, apparently not correlated with other. The mean velocity was 2.2 cms^{-1} (table 2).

Experiment 3 – During this experiment, related to receding level period, the trajectories were very different compared with others experiment. Predominant direction were eastward for the initial 15 h, showing a clear anticyclone circulation after this for drifters 1, 2, 3, 4 and 5 (Figure 7c). The mean velocity was 1.6 cms^{-1} (table 2).

The drifters responded directly to wind-driven current. Moderate current (mean~ 1cms^{-1}) were observed between 20:00 h and 17:00 h (Figure 8 a,b,c,d). Between 17:00 h and 20:00 h a strong intensification in the current was observed with velocities up to 8 cms^{-1}.

Given that the mean wind speed was 1 m s-1 throughout first period (Figure 2c) and 4 ms^{-1} throughout second period, these observations are in accord with those of Hsu (1988) and Curtarelli (2012). The expanding or contracting rate of area was marked by a periodic oscillation (Figure 9). Until typically 16 h after deployment very slow growth was observed. Large divergence events were observed, beginning 16 00 h, 15 00h, 16 00h and 13 30h (Figure 9).

The area increased by a factor of 2 during the Experiment 1 (Figure 9 a,b), by a factor of 6 during the Experiment 2, and by a factor of 3 during the Experiment 3. It is evident from Figure 9 (upper panel) that there was a net growth in the size of the cluster. Such results could have large impacts when estimating the fate of oil-spills or other pollutants. The average value of diffusivity coefficient ($K = 2\ m^2s^{-1}$) is very close to the mean value observed by Stocker and Imberger (2003) ($K = 2.5\ m^2s^{-1}$) in the Lake Kinneret. Comparing the diffusivity coefficient for the identical releases of drifters at the same time but from two slightly different locations (Figure 9 a,b), a pronounced difference is observed. This indicates that the dispersion is sensitive to the location of the release.

Some studies have emphasized the effect of the Earth´s rotation on the lakes response to external disturbances. For example, Antenucci and Imberger (2001) summarized this effect through Burger number ($S = c/Lf$) where c is the nonrotating internal wave phase speed, f the Coriolis frequency, and L the horizontal dimension of the reservoir or lake. These authors indicates that, similarly to conventional Rossby number, when $S \rightarrow 0$, in large lakes and in the ocean, rotation confines the motion to the boundaries, in medium lakes or reservoir it influences the basin-scale response through a balance with stratification (S is O(1)), while in small lakes rotational effects are negligible ($S \rightarrow \infty$). Curtarelli (2011) based on hydrodynamic model and drifters' trajectories indicated the importance of the Earth's rotation to dynamical process occurring in a typical Brazilian reservoir.

Due to increasing knowledgment of the importance of Coriolis parameter to dynamics process occurring in lakes and reservoirs, some process traditionally verified and studied in oceans, should be useful to lakes and reservoir studies. The following section describes one of these processes observed along the Brazil Shelf Break.

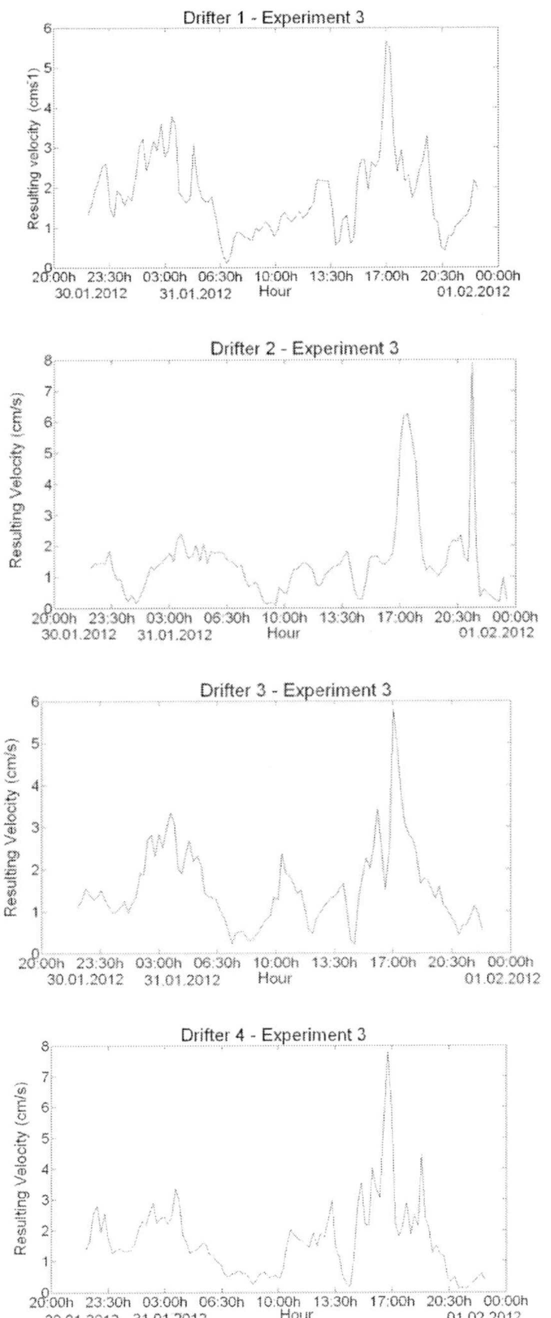

Figure 8. Resulting velocity of the drifters released during the experiment 3.

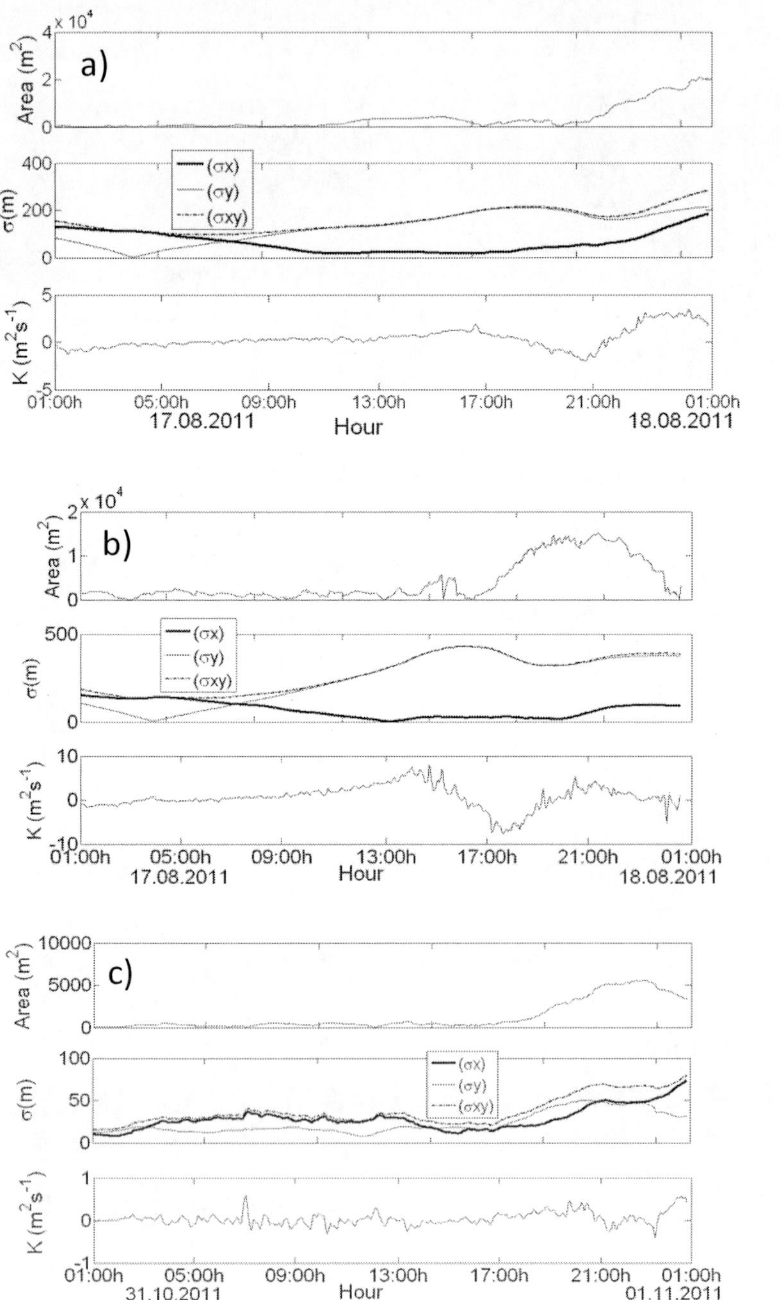

d)

Figure 9. (Continued).

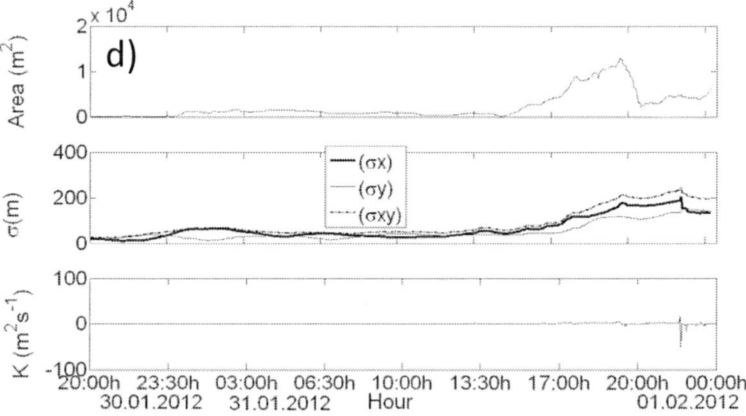

Figure 9. Time evolution of the area formed by drifter's clusters in Experiment 1 (a,b), Experiment 2 (c) and, Experiment 3 (d) (upper panel). Standard deviations σ_x and σ_y along the x and y directions and two-dimensional standard deviation $\sigma_{xy} = (\sigma_x^2 + \sigma_y^2)^{1/2}$ (middle panel). Expanding or contracting rate of a path area on a horizontal plane described by the diffusivity coefficient (K) (bottom panel).

4. Near-Inertial Motions Along the Brazil Shelf Break from Satellite Tracked Drifters

High-frequency horizontal currents, such as inertial and tidal motions, represent an important component of the kinetic energy of the upper oceans (Poulain, 1990). Despite this, little is known about the geographical variability of near-surface inertial currents along the Brazil Current, owing to the difficulty and expense of maintaining moored instruments in the mixed layer and owing to the invisibility of these motions to satellite images.

Satellite tracked drifters represent a remarkable resource useful to study the horizontal variability of inertial currents and its relations with diurnal winds. Recently, 42 GPS positioned drifters were deployed at Brazil coast, under the auspices of MONDO Project (www.prooceano.com.br/mondo). These data are especially useful in the tropics, where the inertial period is around 25 h and therefore is much larger than the Nyquist period of three hours for these drifters.

Some authors have demonstrated that background currents play a critical role for the dynamics of near-inertial motions. The relative vorticity of the eddy field can change the effective Coriolis parameter influencing the near inertial motion (Lerczak *et al.*, 2001), and this can have several results:

generate resonance between the near-inertial currents and the daily wind cycle (Sobarzo *et al.*, 2007); the horizontal spatial scale of the near inertial motions can be set by the horizontal scale of the eddy field (Balmforth *et al.*, 1998); the rate of dispersion of near-inertial energy can be enhanced (Balmforth *et al.*, 1998; Van Meurs, 1998); trapping and amplification of near-inertial motions can occur in regions of intense relative vorticity (Kunze, 1985; D'Asaro, 1995); and spectral broadening of the near-inertial peak can occur due to temporal changes in the background vorticity field (D'Asaro, 1995).

The relationship between inertial currents and meteorological forcing has received increasing attention (Lerczak *et al.*, 2001; Rippeth *et al.* 2002; Sobarzo *et al.* 2007). When diurnal and inertial periods are similar, as occurs in the tropics, analytical models (Hyder *et al.*, 2002) suggest that diurnal winds should drive currents.

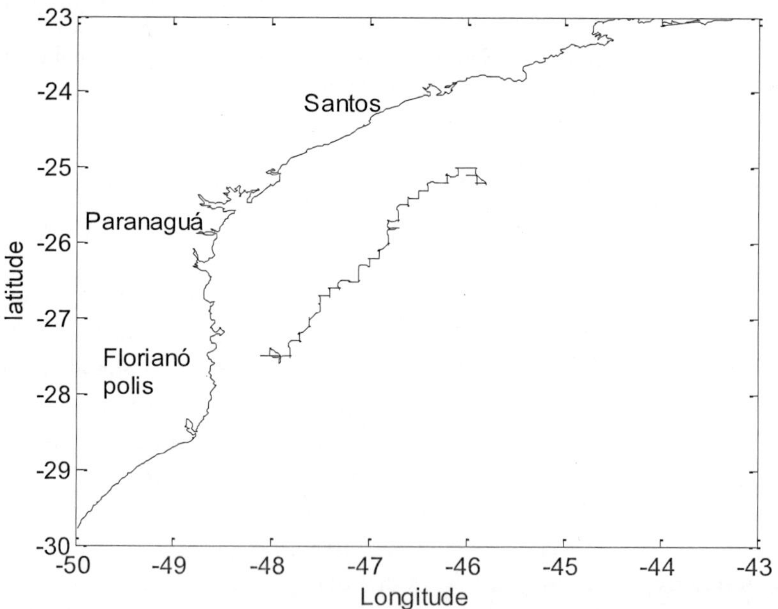

Figure 10. Study area. Drifter trajectory clearly reveals the presence of inertial oscillations.

In this paper will be discussed some characteristics of the inertial currents along the Brazil coast and its relationships with the diurnal wind. A possible cause for the blue shift in frequency of current energy spectra is presented.

4.1. Data Set and Methodology

During the spring of 2007, during the MONDO Project, 40 satellite-tracked drifters were deployed at Santos Basin.

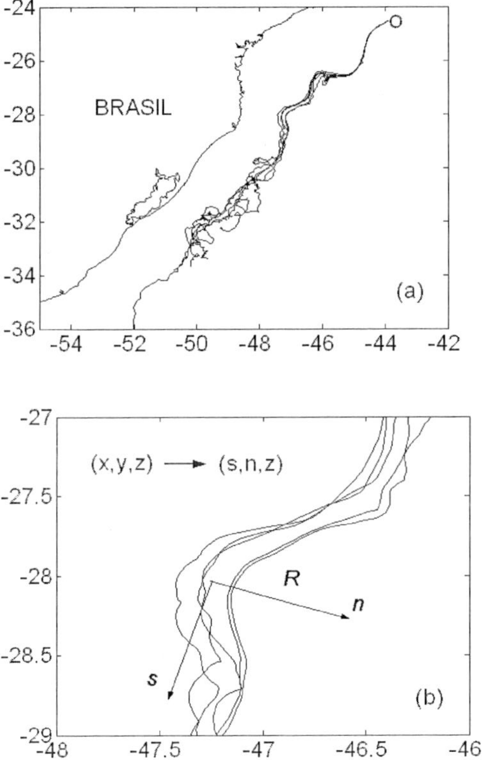

Figure 11. (a) Trajectories and deployment position (empty dot); (b) Natural coordinates systems.

The drifter's positions were obtained by GPS, which allows flow features on the order of 10 m to be resolved. The five trajectories analyzed are shown in Figure 11a.

The vorticity is conveniently determined from the expression in natural coordinates (Arthur, 1965):

$$\xi = \frac{V}{R} - \frac{\partial V}{\partial n}$$

where V is the speed, R is the radius of curvature of the streamline, and is the velocity gradient normal to the streamline (Figure 11b). A tendency for counterclockwise rotation is positive by the convention of signs.

4.2. Wind Data

During spring season the wind blow predominantly from the northeast (Figure 12a) along the Brazilian coast. A long persistence time of northeast wind (Figure 12b) was observed at the period that anteceded the drifter trajectory analyzed here. The intense diurnal variability of wind is indicated (Figure 12c,d).

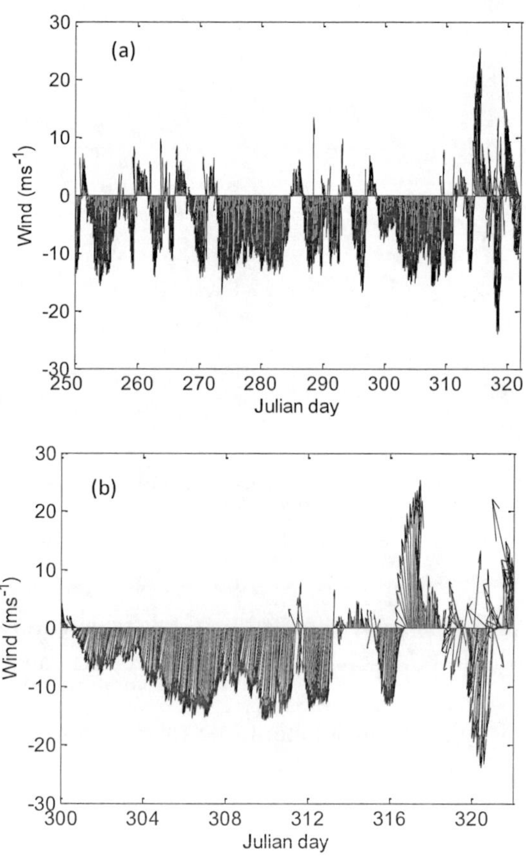

Figure 12. (Continued).

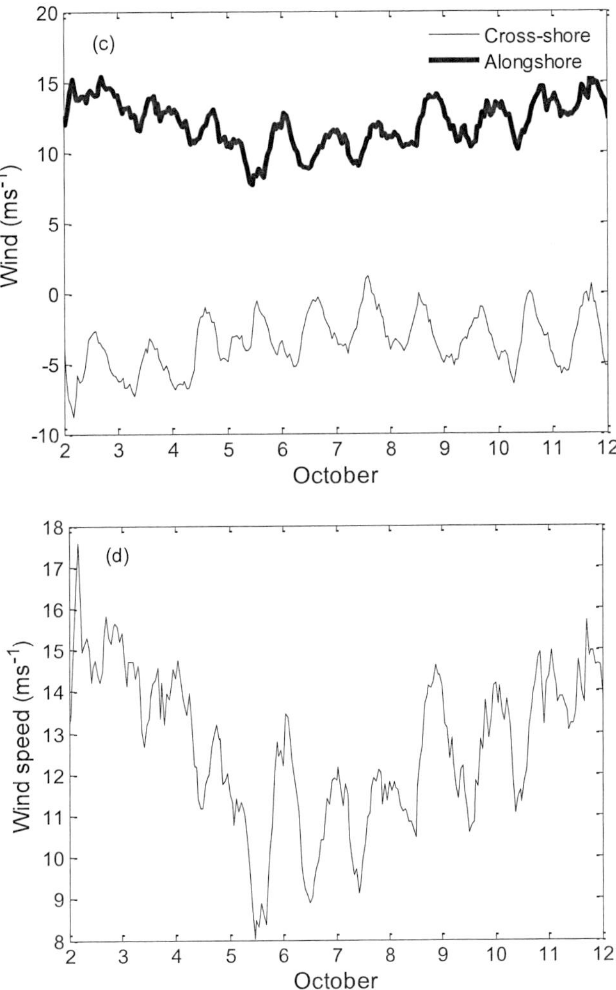

Figure 12. Hourly variation of the horizontal surface wind-vector during spring season (a) and during the drifter trajectory (b); diurnal wind variability (c) and (d).

4.3. Near-Inertial Motions

Following a single drifter track (Figure 10), episodes occur during which the drifter rotates counterclockwise in an almost circular motion. These are near-inertial oscillations with very high amplitudes (up to 60 cm s^{-1}) (Figure 13).

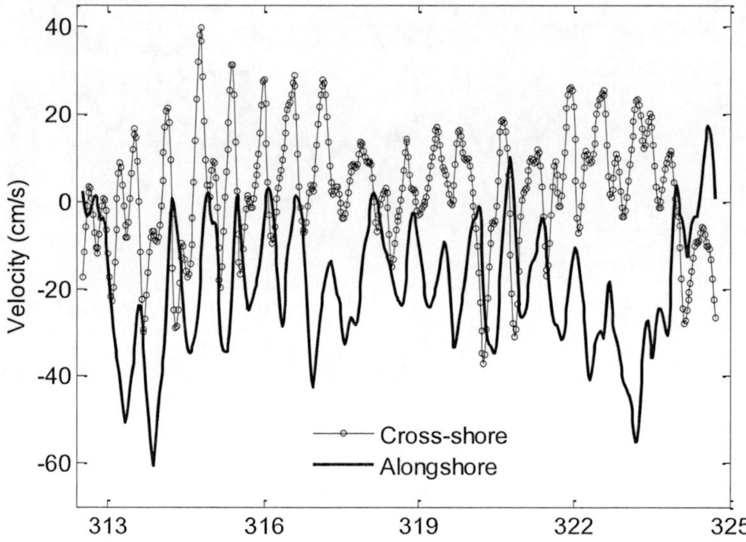

Figure 13. Amplitude of near-inertial currents at 15-m depth (drifters drogued at 15-m depth).

4.4. Discussion

The clear diurnal variation in the wind (Figure 12c,d) is closer the Effective Inertial Frequency (EIF) (~1/25 cph) than the inertial frequency (~1/27cph). The current time series (fig.3) was prevailing counterclockwise, consistent with inertial oscillations to south hemisphere. The percentage of the total variance (Table 3) indicates that 20-25% of the current variance is in the near-inertial band.

Table 3.The percentage of total variance in the near-inertial and superinertial frequency

Período (h)	Percentage of spectral density (%)		
	Cross-shore Alongshore		
25	20		25
12-13	11		17
8-9	8		15

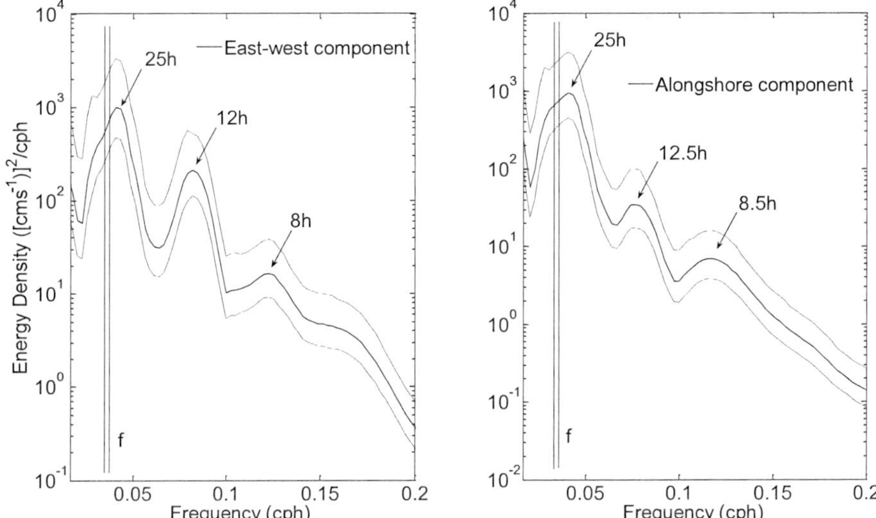

Figure 14. Current spectra. The vertical dashed line indicates the interval for inertial frequency for study area (26-28h).

The spectra analysis (Figure 14) indicated that the near-inertial frequency were about 10% longer than the local inertial frequency. For this occurs in the southern hemisphere (f<0) is required a negative relative vorticity in order to increase the absolute value of the EIF ($|f + \zeta/2|$). This result suggests that the near-inertial motion was embedded in an anticyclonic (negative) vorticity region. The observed near-inertial oscillations with enhanced amplitudes up to 60 cm s^{-1}(Figure 13), would be due to resonance between the diurnal wind and the EIF shifted in direction of diurnal period.

This fact suggests that the near inertial motions can be modulated by the horizontal shear of mean currents and that the relative vorticity is very important to understand the near-inertial movements (where near-inertial is taken to mean near the effective Coriolis frequency) along the Brazil coast. Note that if embedded in a anticyclonic (positive) vorticity field, near-inertial movements have intrinsic frequencies below the inertial or Coriolis frequency which for the latitudes in study area correspond to period between 26 and 28 hs (diverging of diurnal wind and avoiding the resonance possibility).

Other consequence is that the resultant movement has intrinsic frequencies below the effective Coriolis frequency of the surrounding ocean, so cannot propagate out of the region. We speculate that this fact is crucial to explain, despite of the wind favorable conditions, the few near-inertial movements occurrence observed between the 42 drifters launched by the MONDO Project.

The relative vorticity calculated from a set of drifters launched at the same position and time, but for time after the trajectory analysed here, shown the prevailing positive relative vorticity.

The currents observed during MONDO Project were predominantly oriented in the alongshore direction and under potential vorticity conservation (Assireu *et al.*, 2008).

The vorticity of these currents (Figure 13) changed slowly over time and ranged from -1 to 3 x 10^{-5} s^{-1} (0.1 f to 0.5 f). This range was sufficient, at times, to contribute to the effective Coriolis parameter from being subinertial to superinertial and can explain the blue shift (Figure 14).

The 60-cm/s near-inertial current observed at surface (Figure 13) is approximately 4% of the 15 m/s near-steady wind (Figure 12-b), a high value if compared with 2% classical value indicated in literature. It seems as an indication for resonance between the wind and the near-inertial mode of the system.

In order to investigate the resonance mechanism between the wind and the inertial mode of the current, a model of the mixed layer (Pollard and Millard (1970), Poulain (1990), Stech and Lorenzzetti (1992) among others) is used, but under considerations of the Effective Inertial Frequency (EIF),

$$\frac{d\vec{u}(t)}{dt} + 2\pi i[(f + \frac{\zeta}{2})(t) + r]\vec{u}(t) = \frac{\tau(t)}{\rho h} \tag{9}$$

where $\vec{u}(t) = u(t) + iv(t)$ is the horizontal velocity within the mixed layer, $\tau(t) = \tau_x(t) + i\tau_y(t)$ is the wind stress at the surface, f(t) is the local inertial frequency, r is the bottom friction coefficient and h is the water depth of density ρ. Assuming that $\tau = Ae^{iwt}$ and that $V = V_0 e^{iwt}$, the solution of equation (9) is

$$V(t) = \frac{A/\rho h}{[(w + f + \zeta/2)^2 + (r/h)^2]^{1/2}} \exp[i(wt + B)]$$

where B= $\frac{\pi}{2} - \tan^{-1}\left[-\frac{(r/h)}{(w + f + \zeta/2)} \right]$

This solution shows that resonance can occur if $(w+f+\zeta/2)\approx 0$. The dissipation was ignored. This is probably not a problem in the region where the diurnal motions were directly forced by the wind, since most of the energy is near the surface and not subjected to bottom friction. The $(w+f+\zeta/2)$ term only will tend to zero for the southern hemisphere ($f<0$) if $(w+\zeta/2)>0$ and $|f|\cong|w+\zeta/2|$. Thus, the condition for resonance occurrence would be:

$|f|\cong|w+\zeta/2|$ And,

1) For counterclockwise rotation of the wind ($w>0$), $\zeta/2>0$ or $\zeta/2<0$ and $|\zeta/2|<|w|$;
2) For clockwise rotation of the wind ($w<0$), $\zeta/2>0$ and $\zeta/2>w$.

As discussed above, only the cyclonic relative vorticity (negative) is able to shift the EIF higher, towards the diurnal frequency. Then, only the first condition satisfies the request for resonance. Thus, we argue that the condition for resonance is a counterclockwise rotation of the wind ($w>0$) added to negative relative vorticity of current field and that the absolute relative vorticity value are lower than two times the wind frequency. During Julian day between 312 and 324, period of near-inertial movement observed (Figure 13), the wind presented alternate period among northeast and southeast direction (Figure 12b).

As shown by Stech and Lorenzzetti (1992), the wind in the Brazilian coast rotate counterclockwise from northeast to southeast ($w<0$) with the frontal passages.

Considering a diurnal wind variability ($w\sim 1/(24 \times 3600) \sim 1.16 \times 10^{-5}\ s^{-1}$) and considering that the relative vorticity presented in Figure 15 can be assumed as a typical value ($-1.0 \times 10^{-5}\ s^{-1}$), then $|\zeta/2|<|w|$. Thus, under hypothesis of negative relative vorticity, we arrive in results like the Lerczak *et al.* (2001) and Sobarzo *et al.* (2007), that the large near diurnal current response in the upper ocean (Figure 13) is associated with the resonance with the diurnal wind forcing.

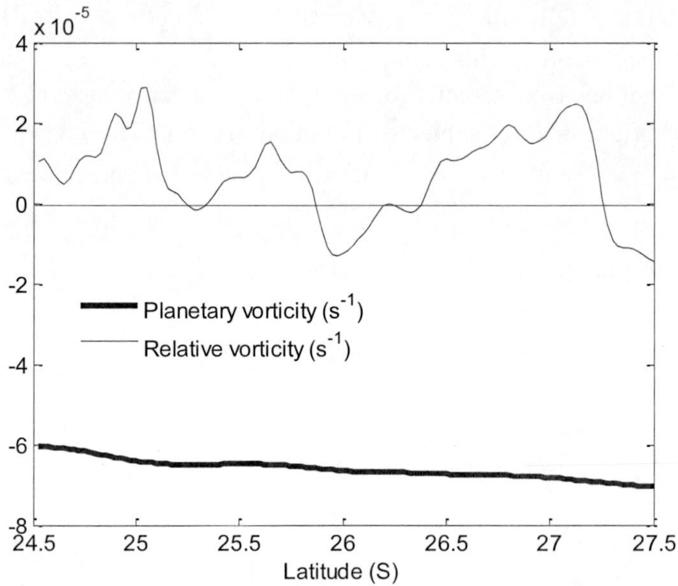

Figure 15. Relative and planetary vorticity.

Conclusion

Compared with tracer, drifter has the advantages and disadvantages common to most remote-sensing techniques. The main disadvantage is that we are not measuring concentration directly, so the spreading due to vertical shear dispersion can introduce errors. However, as previously shown by some authors, the vertical shear dispersion is small compared to horizontal shear dispersion.

Then, the drifter´s motions are representative of the behavior of diffusive substances. The main advantages of drifters are the greater experimental flexibility and the accuracy due to improvements in positioning technology over recent years. The diffusivity observed in this study is comparable with those observed in others lakes. Comparisons between the diffusivity coefficient for the identical releases of drifters at the same time but from two slightly different locations showed a pronounced difference. This indicates that the dispersion is sensitive to the location of the release.

Poleward-propagating, near-inertial motions with amplitude up to 60 cm s^{-1} dominated a drifter trajectory along the Brazilian Shelf Break (BSB). This near-inertial current variability accounts for 20-25% of the total observed current variance. The background current vorticity field showed relevant role over the near-inertial motions on the Brazil Shelf Break. As an example, if embedded in a anticyclonic (positive) vorticity field, near-inertial movements have intrinsic frequencies below the inertial or Coriolis frequency which for the latitudes in study area correspond to period between 26 and 28 hs (diverging of diurnal wind and prevent the resonance possibility).

The near-inertial frequency were about 10% longer than the local inertial frequency corresponding in a shift towards the diurnal frequency, what generated conditions to resonance between the near-inertial currents and the daily wind cycle, enhancing the near-inertial amplitude. Results from analytical model indicated that the conditions for BSB resonance with the diurnal wind forcing based on a counterclockwise rotation of the wind added to negative relative vorticity of current field and that the absolute relative vorticity value are lower than two times the wind frequency.

Acknowledgments

The authors want to thank FAPEMIG, the research supporting foundation for the State of Minas Gerais (Project APQ – 02582-10). This study would not have been possible without the assistance of Daniel Mortl, who passed away in early February of 2012.

References

Antenucci,J.P., Imberger, J. (2001). Energetics of long in-terval gravity waves in large lakes. *Limnol. Oceanogr.* 46:1760–1773.

Balmforth, N.J., Llewellyn Smith, S.G., Young, W.R. (1998). Enhanced dispersion of near-inertial waves in an idealized geostrophic flow. J. *Mar. Res.*, 56, 1-40.

Corell H., Moksnes P-O., Engqvist A., Döös K. and Jonsson P.R. (2011). Larval traits determine dispersal distance and optimum size of marine protected areas. *Marine Ecology Progress Series* (in press).

Curtarelli, M. P. (2012) Estudo da influência de frentes frias sobre a circulação e os processos de estratificação e mistura no reservatório de Itumbiara (GO): um enfoque por modelagem hidrodinâmica e sensoriamento remoto. 108 p. Dissertação de mestrado em Sensoriamento Remoto - Instituto Nacional de Pesquisas Espaciais, São José dos Campos.

D'Asaro, E.A., Eriksen, C.C., Levine, M.D., Niiler, P., Paulson, C.A., Van Maurs, P. (1995) Upper-Ocean inertial currents forced by a strong storm. Part I: Data and comparisons with linear theory. *J. Phys. Oceanogr.*, 25, 2909-2936.

HSU, S. A (1988). Coastal Meteorology. San Diego: Academic Press, 260 p.

Hyder, P., Simpson, J.H., Christopoulos, S., (2002). Sea-breeze forced diurnal surface currents in the Thermaikos Gulf, North-west Aegean. *Cont. Shelf Res.* 22, 585–601.

Johnson, D., Stocker, R., Head, R., Imberger, J., Pattiaratchi, C. (2003). A compact, low-cost GPS drifter for use in the oceanic nearshore zone, lakes, and estuaries. *Journal of Atmospheric and Oceanic Technology*, 20, 1880 – 1886.

Kunze, E.L., (1985). Near-inertial waves propagation in geostrophic shear. J. *Phys. Oceanogr.* 15, 544–565.

Lawrence, G. A., Shley, K. I. A, Onemitsu, N.Y, Llis, J. R. E. (1995). Natural dispersion in a small lake. *Limnol. Oceanogr*. 40: 1519-1526.

Lekien, F., Coulliette, C., Mariano, A. J., Ryan, E. H., Shay, L. K., Haller G., Marsden J.(2005) Pollution release tied to invariant manifolds: A case study for the coast of Florida. *Physica D* 210 (1-2), 1–20.

Lerczak, J.A., Hendershott, M.C., Winant, C.D., (2001). Observations and modelling of coastal internal waves driven by a diurnal sea breeze. *J. Geophys. Res.* 106 (C9): 19715–19729.

Macintyre, S.; Romero, J.R.; Kling, G.W (2002). Spatial-temporal variability in surface layer deepening and lateral advection in a embayment of Lake Victoria, East Africa, *Limnology and Oceanography*, v. 47, p. 656-671.

Marcus Sobarzo, R. Kipp Shearman, Steve Lentz. (2007). Near-inertial motions over the continental shelfoff Concepcio´ n, central Chile. *Progress in Oceanography* 75:348–362.

Molinari, R.; Kirwan, A.D.J.(1975). Calculations of differential kinematic properties from Lagrangian observations in the Western Caribbean Sea. *J. Phys. Oceanogr.* 5:483–491.

Niiler, P. P., Sybrandy, A. S., Bi, K., Poulain, P-M., Bitterman, D. (1995) Measurements of the water-following characteristics of Tristar and Holey-sock drifters. *Deep Sea Research*, 42:1951 – 1964.

Okubo, A.; Ebbesmeyer, C.C..(1976). Determination of vorticity, divergence and deformation rates from analysis of drogue observations. *DeepSea Res.* 23:345–352.

Okubo, A.; Ebbesmeyer, C.C.; Helseth, J.M. (1976) Determination of Lagrangian Deformations from Analysis of Current Followers. Journal of *Physical Oceanography*, 6: 524-527.

Pacheco, F.S.; Assireu, A.T.; Roland, F. (2011) Derivadores rastreados por satélite aplicados a ambientes aquáticos continentais: caso do Reservatório de Manso. In: ALCÂNTARA, E.H.; STECH, J.L.; NOVO, E.M.L.M. (orgs.). Novas Tecnologias para o Monitoramento e Estudo de Reservatórios Hidrelétricos e Grandes Lagos. Rio de Janeiro: Parêntese Editora, p. 193-218. ISBN 9788560507108.

Poulain, P-M. (1990) Near-inertial and diurnal motions in the trajectories of mixed layer drifters. *Journal of Marine Research*, 48:793-823.

Prahl, F.G., Crecellus, E., Carpenter, R. (1984) Polycyclic aromatic hydrocarbons in Washington coastal sediments: an evaluation of atmospheric and riverine routes of introduction, *Environ. Sci. Technol.* 18:687–693.

Rice, D.W., Seltenrich, C.P., Spies, R.B., Keller M.L. (1993) Seasonal and annual distribution of organic contaminants in marine sed-iments from the elkhorn slough, moss landing harbor and near-shore Monterey Bay, California, *Environ. Pollut.* 82:79–91.

Rippeth, T.P., Simpson, J.H., Player, R.J., Garcia, M., (2002). Current oscillations in the diurnal–inertial band on the Catalonian shelf in spring. *Cont. Shelf Res.* 22: 247–265.

Sanderson, B. G. (1995). Structure of an eddy measured with drifters. *J. Geophys. Res.* 100: 6761–6776.

Sanderson, B. G., Pal, B. K., Goulding, A. (1988) Calculations of unbiased estimates of the magnitude of residual velocities from a small number of drogue trajectories. *Journal of Geophysical Research*, 91(C7):8161 – 8162.

Soomere T., Viikmae B., Delpeche N., and Myrberg K. (2010). Towards identication of areas of reduced risk in the Gulf of Finland, the Baltic Sea. *Proceedings of the Estonian Academy of Sciences* 59:156-165.

Stoker, R.; Imberger, J (2003). Horizontal transport and dispersion in the surface layer of a medium size lake. *Limnology and Oceanography*, 48: 971-982.

Van Meurs, P. (1998). Interaction between near-inertial mixed layer currents and the mesoscale: the importance of spatial variability in the vorticity field. *J. Phys. Oceanogr.*, 28:1363-1388.

In: Remote Sensing
Editor: Enner Alcântara
ISBN: 978-1-62417-140-6

Chapter 5

REMOTE SENSING AND ATTITUDE CONTROL USING INTELLIGENT IMAGE-BASED SENSOR

Gregor Klančar**, Drago Matko, Gašper Mušič and Sašo Blažič

Centre of Excellence for Space Sciences and Technologies, Ljubljana, Slovenia
University of Ljubljana, Faculty of Electrical Engineering, Ljubljana, Slovenia

ABSTRACT

Remote sensing using low-Earth orbit satellites usually needs navigation sensors as well as imaging sensors to perform required Earth observation tasks. The main purpose of the presented research is to use a visual sensor, in this case an imaging camera, as an intelligent device for remote sensing and for extracting relative attitude information needed for the attitude control. Natural features are identified from the camera image to recognize and track the desired observation scene. Based on the estimated tracking error, the image-based control law is proposed to achieve an automatic scene tracking, which produces the required moments around satellite axes. The proposed image-based control is verified using the satellite simulator realized in the Matlab environment. The position of the satellite is simulated with the Simplified General Perturbations model and its orientation by simulating its dynamic and

* E-mail address: {gregor.klancar, drago.matko, gasper.music, saso.blazic}@fe.uni-lj.si

kinematic models. From the satellite pose and the line of sight of the camera, the Earth's surface image is obtained from the Google Earth application, which simulates the satellite camera. The results of the simulations performed by virtual satellites and performance analysis obtained from real satellites' data show the approach applicability. Main contributions of the presented chapter are summarized as follows: simplicity of the approach due to the same sensor used for attitude control and Earth remote sensing, good tracking results with no special explicit calibration of camera parameters being required.

Keywords: satellite, remote sensing, attitude control, computer vision, AMS Subject Classification: 34H, 65D, 00A

1. Introduction

Earth observation from satellites in low orbits is an important task, not only for military use but also for environmental monitoring, meteorology, natural disaster monitoring, agriculture, land cover cartography and the like. For that purpose, smaller satellites, equipped with visual camera, are also used, which, due to lower required budget and shorter development times, enable smaller players such as universities to obtain space technology. Such affordable missions offer the opportunity for different organizations to target specific applications and provide emerging space nations with independent Earth-observation facilities (Wicks et al., 2000).

Earth-observation satellites need an attitude-control system to steer the satellite to the desired area on the Earth's surface for observation purposes or to utilize a broadband downlink. Usually, automatic attitude control is used, where the operator only controls the desired target area, while the satellites track it autonomously. In most applications, Earth target tracking is solved using an extended Kalman filter in an attitude-localization unit, such as in Steyn (2006) and Psiaki et al. (1990), where estimates of the satellite attitude are obtained using relative and absolute sensor information. Such applications require navigation sensors (Earth magnetic, gyros, sun sensor and the like) and observation sensors (e.g., a camera). In these approaches, information from the camera is not used in the satellite attitude control. However, there are some examples of satellite camera usage for automation purposes, such as satellite image registration or region-of-interest detection.

The use of satellite camera image information for an automatic satellite image registration process was reported by Wong and Clausi (2009) and Fan et al. (2007), where feature pairs from sensed and reference images are identified. The compression of high-resolution JPEG satellite images using region-of-interest detection and coding by Fuzzy C-Means clustering is presented in Raj et al. (2008).

A number of successful approaches in UAV (Unmanned Aerial Vehicle) localization and control (Škrjanc, 2006) were reported using camera and natural landmark information. In Cesetti et al. (2010), an improved version of the Scale Invariant Feature Transform (SIFT) is used for the navigation and safe landing of a helicopter. In Erhard et al. (2010), an analysis of different feature-extraction algorithms for the localization purposes of a quadrocopter are proposed. Vision-only navigation and the control of a small, autonomous helicopter using a Sigma-Point Kalman Filter are proposed in Bai et al. (2007).

The main purpose of our current research is to use a visual sensor, in this case an imaging camera, as an intelligent device extracting relative attitude information needed for attitude control. The same camera is used for observation as well as for attitude control. Satellite orientation error is estimated and corrected by a proposed computer vision and visual servoing algorithm. The presented work is a continuation of our previous research with some results already published in Klančar et al. (2011). The proposed approach has some important advantages, as follows: simplicity of the approach due to the use of the same sensor for observation and attitude control, no need for explicit calibration of camera parameters and better tracking accuracy (compared to external attitude sensors' performances).

Considering the above-mentioned advantages, the proposed idea is especially attractive for small-satellite missions. The satellite achieves the desired Earth-observation spot using image-based control (visual servoing), which together with an onboard camera forms an attitude control loop. The proposed control utilizes only natural features (Scale Invariant Feature Transform – SIFT, Lowe, 2004) to recognize the desired observation scene, which to the best of our knowledge is the first such approach to satellite attitude control.

In general, visual servoing techniques are classified into two main types: image-based control and position-based control. Image-based control uses features that are extracted from the image to directly provide control actions for actuators, as opposed to position-based control where the pose (attitude) of the object is firstly estimated from the image and then issued to pose-based control. In our case, image-based control is more convenient to achieve a constant Earth-area observation because the control goal is not to directly achieve the reference satellite orientation given for all three Euler angles but to control satellite orientation so that the desired Earth area is always in the image center. Therefore, features calculated from each frame are compared to the features obtained from the reference images.

The reference image is selected to cover the desired observation area on the Earth. By comparing features from the current image with the features from the reference image, the translation and rotation of the current image relative to the reference image are estimated. To observe the desired scene, the translation error can be compensated by the roll and pitch angle of the satellite and the orientation error by the yaw angle.

This is achieved by the proposed control structure, which according to the errors in image translation and orientation produces the required moments around satellite axes to cancel the orientation error. In this work, the mathematical models for determining the satellite position and orientation are implemented in the Matlab environment. They are used to simulate satellite motion in Earth orbit. From a known satellite position and orientation, the satellite camera is simulated using the Google Earth application that provides camera images. Finally, these images are used to close the control loop with the implemented image-based control.

The chapter is organized as follows: In Section 2, the mathematical models of the position of the satellite, its orientation and camera focus point are given. The attitude of the satellite is governed by the image-based control law described in Section 3. The simulation results are presented in Section 4, and the conclusions are drawn at the end.

2. Mathematical Models

Mathematical models are needed to simulate a satellite's motion in Earth orbit (position and rotation) and its onboard camera operation. These models are a part of the simulation environment developed in this work.

In the following, the mathematical models of the satellite's position, its orientation and camera focus point will be given. For the satellite position, the Simplified General Perturbations (SGP) model (Hoots and Roehrich, 1988) is used. The orientation of the satellite is determined by its dynamic and kinematic models, while the calculation of the camera's focus point is a geometric problem. All these models are implemented in a Matlab-based satellite motion simulator.

2.1. SGP Model

The Simplified General Perturbations model is used to simulate the satellite's position in its orbit, which is defined by the Kepler elements that describe the satellite's position at the initial time. The satellite's position at some arbitrary time can then be calculated as follows.

Given the Kepler elements of the orbit (Wertz, 1978):

- i – inclination,
- Ω_0 – Right Ascension of Ascending Node (RAAN) at epoch,
- ω_0 – argument of perigee at epoch,
- M_0 – mean anomaly at epoch,
- ε – eccentricity,
- n – mean motion

obtained from the NORAD two-line element set (Wertz, 1978), where the epoch (initial time) t_0 and $\frac{1}{2}\frac{dn}{dt}$ are also given, the position of the satellite is calculated as follows:

1. Calculate the mean anomaly M at time

$$M = M_0 + n(t - t_0) + \frac{1}{2}\frac{dn}{dt}(t - t_0)^2 \tag{1}$$

2. Calculate (by iterations) the eccentric anomaly from the mean anomaly using the Kepler equation

$$M = E - \varepsilon \sin(E) \tag{2}$$

3. Calculate the true anomaly φ

$$\varphi = \arctan\left[\frac{\sin(E)\sqrt{1-\varepsilon^2}}{\cos(E) - \varepsilon}\right] \tag{3}$$

4. Calculate the semi-major-axis a from the mean motion

$$a = \sqrt[3]{\frac{\mu}{n^2}} \qquad \mu = 3.986005 \cdot 10^{14} \tag{4}$$

5. Calculate the actual argument of perigee ω and RAAN Ω due to the geopotential coefficient $J_2 = 1.0826310^{-3}$ (Hoots and Roehrich, 1988)

$$\omega = \omega_0 + \frac{d\omega}{dt}(t - t_0) \tag{5}$$

$$\Omega = \Omega_0 + \frac{d\Omega}{dt}(t - t_0) \tag{6}$$

where,

$$\frac{d\omega}{dt} = \frac{3}{4} n (\frac{a_E}{a})^2 \frac{5\cos^2 i - 1}{(1-\varepsilon^2)^2} J_2 \tag{7}$$

$$\frac{d\Omega}{dt} = -\frac{3}{2} n (\frac{a_E}{a})^2 \frac{\cos i}{(1-\varepsilon^2)^2} J_2 \tag{8}$$

and $a_E = 6378 \cdot 10^3$ is the semi-major-axis of the Earth ellipsoid.

6. Calculate the position of the satellite in the Earth Centered Orbit (ECO) frame

$$P_{ECO} = \begin{bmatrix} \dfrac{a(1-\varepsilon^2)\cos(\varphi)}{1+\varepsilon^2\cos(\varphi)} \\ \dfrac{a(1-\varepsilon^2)\sin(\varphi)}{1+\varepsilon^2\cos(\varphi)} \\ 0 \end{bmatrix} \tag{9}$$

7. Transform the ECO position into the position of the satellite in the Earth Centered Inertial (ECI) frame

$$P_{ECI} = R_{ECO}^{ECI} P_{ECO} \tag{10}$$

where,

$$R_{ECO}^{ECI} =$$

$$= \begin{bmatrix} c(-\Omega) & s(-\Omega) & 0 \\ -s(-\Omega) & c(-\Omega) & 0 \\ 0 & 0 & 1 \end{bmatrix} \begin{bmatrix} 1 & 0 & 0 \\ 0 & c(-i) & s(-i) \\ 0 & -s(-i) & c(-i) \end{bmatrix} \begin{bmatrix} c(-\omega) & s(-\omega) & 0 \\ -s(-\omega) & c(-\omega) & 0 \\ 0 & 0 & 1 \end{bmatrix}$$

2.2. Dynamic and Kinematic Models of the Satellite

The dynamic and kinematic models of the satellite are as follows (Wertz, 1978):

$$\frac{d\omega}{dt} = J^{-1}(M - \omega \times (J\omega)) \tag{11}$$

where J is the tensor of the satellite's moments of inertia, $\omega = [\omega_u \quad \omega_v \quad \omega_w]^T$ is the vector of the angular velocities of the satellite with respect to the ECI frame (but expressed in the satellite frame), M are the moments applied to the satellite (expressed in the satellite frame) and $\times$ denotes the vector product.

As for kinematical model, we have, due to the simplicity of calculation, chosen the description with the Direction Cosine Matrix (DCM) R_{ECI}^{Body}

$$\frac{dR_{ECI}^{Body}}{dt} = \begin{bmatrix} 0 & \omega_w & -\omega_v \\ -\omega_w & 0 & \omega_u \\ \omega_v & -\omega_u & 0 \end{bmatrix} R_{ECI}^{Body} \tag{12}$$

2.3. Calculation of the Focus Point

The line of sight of the camera is oriented in the z direction of the satellite coordinate system. The focus point on the surface of the Earth in the ECI coordinate system

$$F_{ECI} = \begin{bmatrix} x_f \\ y_f \\ z_f \end{bmatrix} \tag{13}$$

is then calculated from the intersection of the Earth's surface (it is supposed that the Earth is a sphere) and the line of sight

$$\begin{aligned} &x_f^2 + y_f^2 + z_f^2 = R_E^2 \\ &P_{ECI} - Dd = F_{ECI} \end{aligned} \tag{14}$$

where R_E is the Earth-sphere radius, D is the unit vector of the direction of the line of sight of the camera and d is the distance from the satellite position to the focus point. The system of four equations (14) with four unknowns (x_f, y_f, z_f, d) is solved analytically using the symbolic toolbox of Matlab. If the line of sight of the camera is in the direction of the Earth, the system has two real solutions, and the nearest to the satellite (smallest d) is used. In each step, the actual position of the satellite (P_{ECI}) and the actual unit vector of the direction of the line of sight of the camera (oriented in the z direction of the Body coordinate system) in the ECI coordinate system

$$D = R_{ECI}^{Body^{-1}} \begin{bmatrix} 0 \\ 0 \\ 1 \end{bmatrix} \tag{15}$$

are used to calculate F_{ECI}. The focus point on the Earth in the Earth Centered Earth Fixed coordinate system (ECEF) is then calculated using

$$F_{ECEF} = \begin{bmatrix} \cos(\omega_E t) & \sin(\omega_E t) & 0 \\ -\sin(\omega_E t) & \cos(\omega_E t) & 0 \\ 0 & 0 & 1 \end{bmatrix} F_{ECI} \quad (16)$$

where t is the Greenwich Siderial Time and $\omega_E = 7.2921 \cdot 10^{-5}$ is the rotational rate of the Earth.

3. Image-Based Control of the Satellite

Control of the satellite's orientation is performed using image-based (IB) control, where the simulated camera images are obtained from the Google Earth application. Successive images are compared based on extracted SIFT (Scale Invariant Feature Transform) features developed by Lowe (2004). The SIFT is convenient for image-feature generation in object-recognition applications (Se and Little, 2002) because they are invariant with respect to image translation, scaling, rotation, and partially invariant to illumination changes and affine or 3D projection.

In the proposed approach, at each frame, the SIFT features are calculated and compared to the SIFT features from the reference image. The reference image is selected to cover the desired small area on the Earth that we want to continually observe. By comparing features from the current image with the features from the reference image, the required control actions (reaction wheels) are determined to reorient the satellite to the reference focus point on the Earth's surface.

3.1. Capturing Images from Google Earth

We used Google Earth as an Earth's surface image generator and to simulate the satellite camera. The application has to be controlled from Matlab to reproduce the satellite view of the Earth's surface for a calculated satellite position and orientation. To achieve this, Matlab connects to the ActiveX server interface of Google Earth and then uses a part of the available properties and methods. Among others, the interface to the camera is available, which can be used to set the focus point as well as the range, tilt and azimuth of the camera.

These properties are repeatedly set to match the desired focus point and satellite position during the simulation experiment. Once a proper image is displayed within the Google Earth render window, the image is captured by a call to a Matlab mex function written in C. Finally, the acquired image is rotated to match the satellite orientation.

3.2. SIFT Features Generation

The SIFT feature locations in the image are calculated by comparing a sequence of the same, but scaled, images as follows (Se and Little, 2002; Lowe, 2004). The image is sequentially filtered with the Gaussian kernel. The filtered images are subtracted from among each other to obtain the sequence of difference-of-Gaussian images.

The smoothed images are then down-sampled, and a new sequence of difference-of-Gaussian images is computed. This process is repeated until the image is too small for a reliable feature detection.

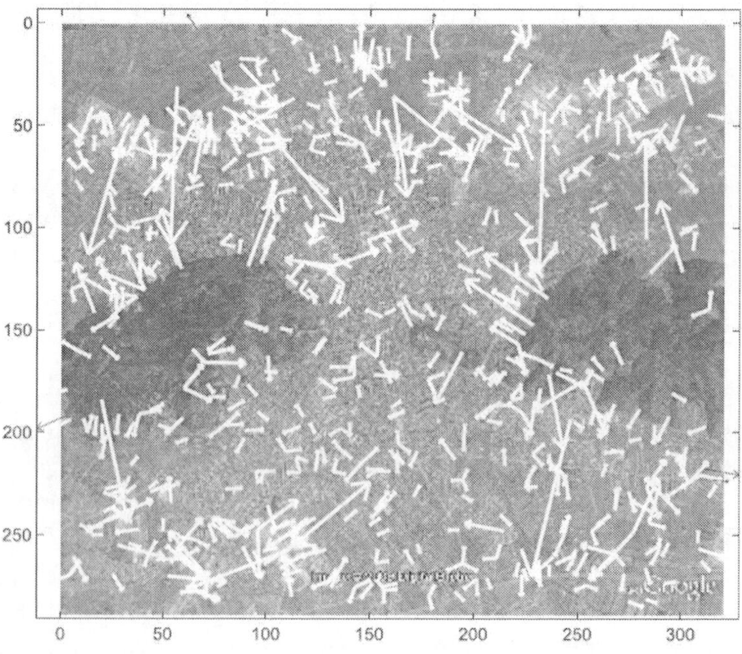

Figure 1. SIFT features (arrows) found on the observed Earth image (in the pixels image 729 features are found).

Within the difference-of-Gaussian pyramid, the maxima and minima for each sample point are detected by comparing the neighbors in the current image and the neighbors in the scale above and below. This procedure gives stable features according to the translation, scale, rotation and illumination conditions.

In this work, the SIFT features were calculated using the toolbox (Lowe, 2005), where each feature is represented by a descriptor vector with 128 elements and by a four-element location vector. An example of the found SIFT features for a typical Google Earth image is given in Figure 1. Each feature is shown with the location vector defined by the row and column image coordinates the scale (vector size) and the orientation.

3.3. Feature Matching

For each sample time (defined by the camera frame rate), the current image features are matched to the reference image features so that each found pair of features represents the same point on the Earth's surface. The pairs are found by searching the feature space (descriptor). For each feature d in the current image, a pair from the reference image d_r is found by searching for a minimal distance between them. To reduce the computational costs in Matlab, this procedure is done by calculating the dot product of the normalized features ($|d|=|d_r|=1$) as follows: $d \cdot d_r = \cos\beta$, where β is the angle between the feature vectors d and d_r. For small angles β, the distance among the features can be approximated by:

$$\left|d-d_r\right| = D \approx \beta = \arccos\left(d \cdot d_r\right) \tag{17}$$

To increase the robustness, only pairs fulfilling the condition $\frac{D}{D_{sec}} < k$ are considered, where D isthe distance among the features in the found pair, D_{sec} is the distance of the current feature to the second-nearest feature from the reference image and $0<k<1$ ($k=0.6$). This means that the feature pairs are valid if no other feature is in the vicinity.

3.4. Image Feature Detectors Performance Analysis

Satellite attitude control is done based on the information obtained from the camera images. Image association problem can be solved by applying different feature detectors. For such a task, many feature detectors have been proposed, and also some detailed comparisons of feature detectors were performed in Tuytelaars and Mikolajczyk (2007; CVT, 2011). In general, the desired feature properties are repeatability (between two different views of the same scene), quantity (a reasonable number of features for selected scene), computational efficiency (real-time operation), robustness (to image translation, rotation, scaling and illumination) and other properties (Tuytelaars and Mikolajczyk, 2007).

However, the proper choice of feature detector depends on a problem. In the following, the performance analysis of some feature detectors implemented in OpenCV library (SIFT, SURF, FAST, STAR, MSER, GoodFeaturesToTrack (GFTT)) is made. For this analysis, a Matlab toolbox coded by Zdešar (2012) is used. Comparison is done on real satellite images of Earth surface such as urban areas, landscapes, sea and areas partly covered by clouds (see Figure 2).

Performance analysis of different feature detectors calculated from images in Figure 2 is summarized in Figs. 3-6, where numbers 1 to 7 on the x axis correspond to the images in the rows in Figure 2. In Figure 3, the average number of found features (from the first and the second image in each row) is displayed for each feature detector.

The number of feature pairs found using the distance criteria (17) are given in Figure 4. From all found pairs, the percentage of good feature pairs are given in Figure 5.

Good pairs are computed using RANSAC method where mathematical motion model is fitted to the features, and outlier features, which cannot be fitted to this model, are eliminated. The computational time needed to compute features from an image is given in Figure 6.

The number of found features or feature pairs (see Figures 3, 4) is not the measure of feature detector quality. More important is the number of good features or pairs (see Figure 5), which can be reliable used in the image-based control algorithm proposed in the following. The appropriate feature detector selection depends on the problem specification, so an analysis such as presented here should be performed. From the presented comparison, it follows that SIFT and SURF features are the most reliable in all the image scenarios given in Figure 2.

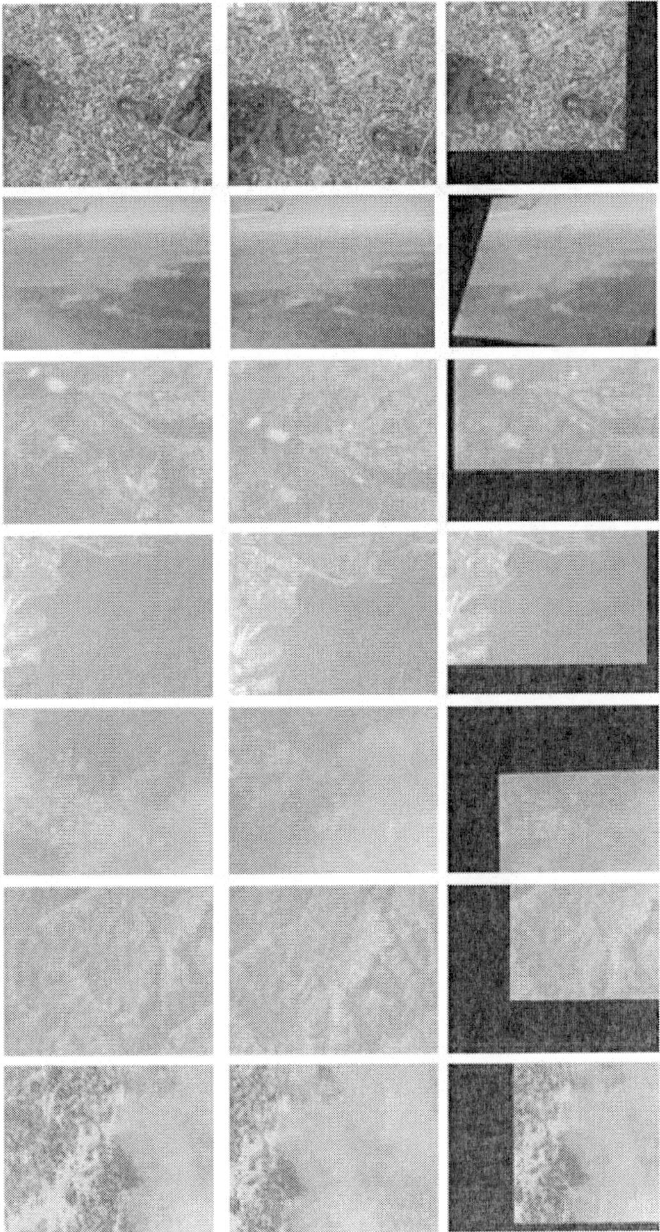

Figure 2. Images used for feature detectors comparison. Columns: first image, second image and alignment. Rows: simulated camera images, images taken from plane and real satellite images.

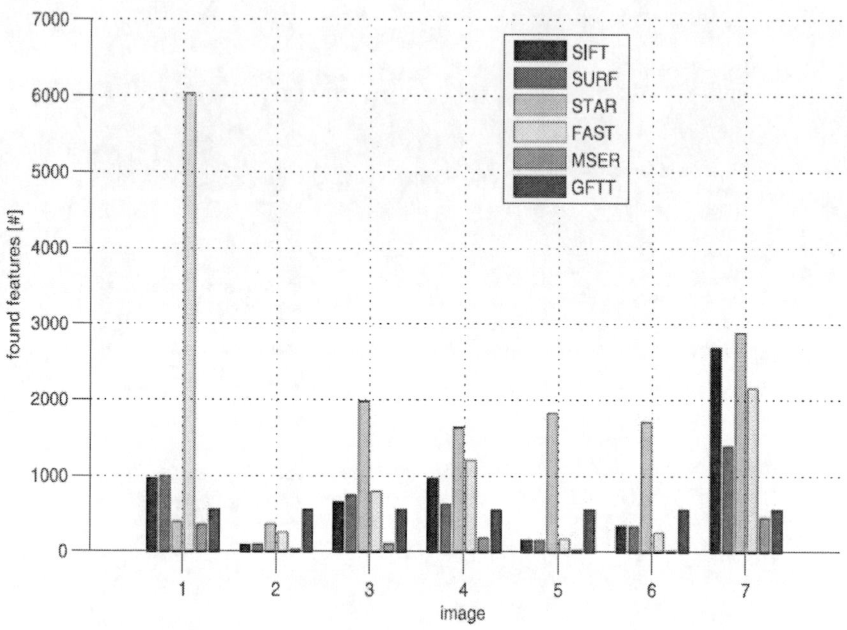

Figure 3. The number of found features for each feature detector.

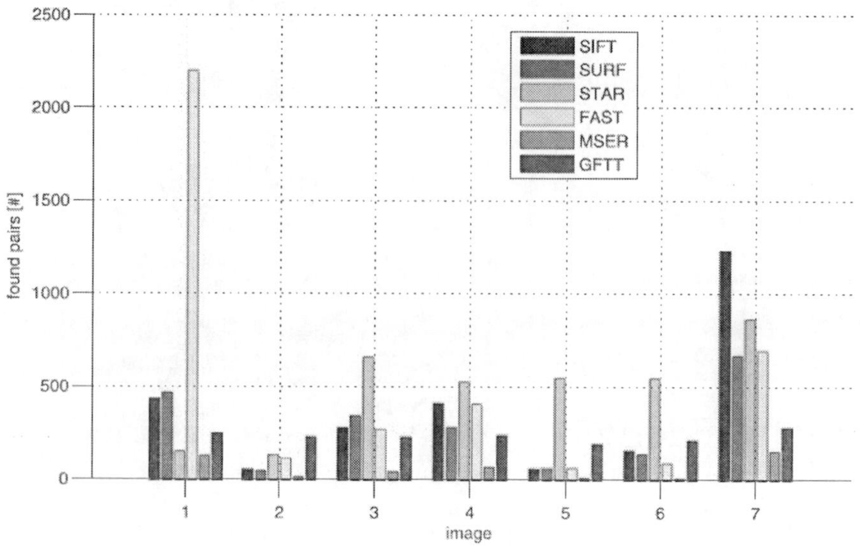

Figure 4. The number of feature pairs found based on distance criteria (Wertz , 1978).

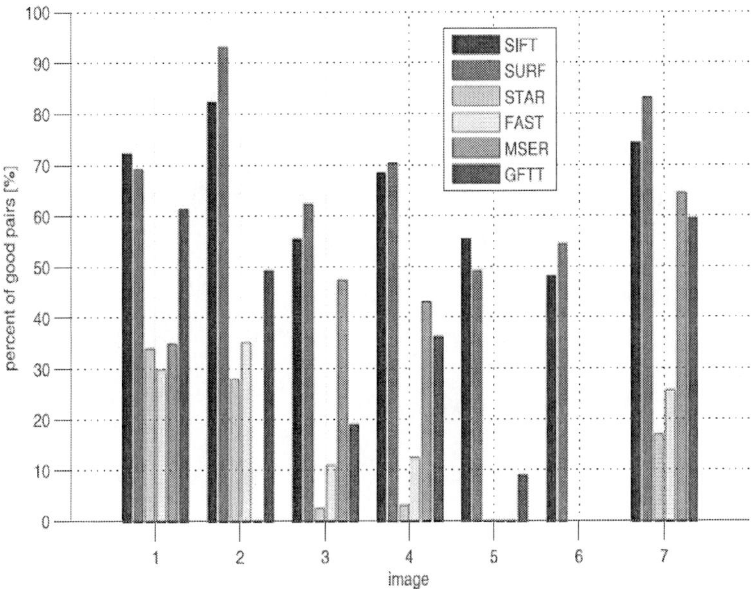

Figure 5. The percentage of good feature pairs computed using RANSAC method.

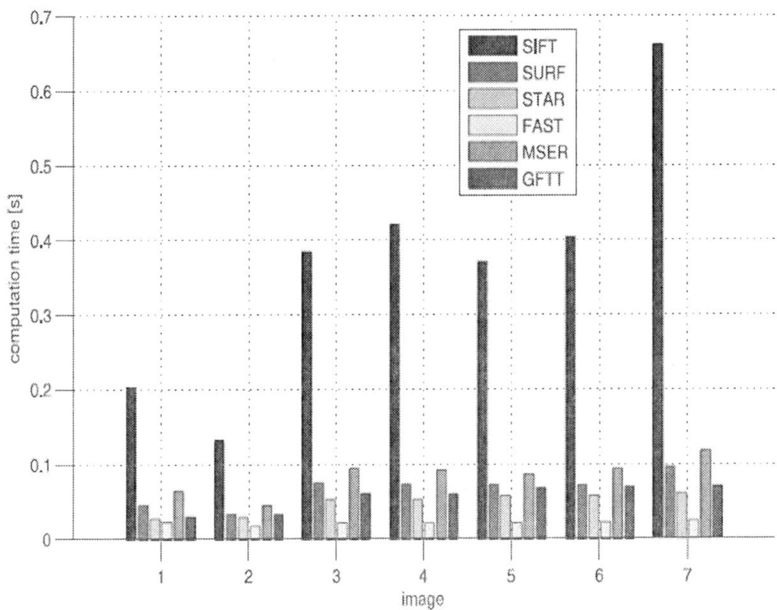

Figure 6. The computational time needed to compute different features.

Other features give good results only on some images and fail to produce reliable features in other images. The performed comparison also confirms that the simulated images are comparable to the true satellite images and can, therefore, be used to design and validate the proposed image-based attitude control algorithm. The computational time depends on algorithm complexity and also on implementation of the algorithm. SIFT features need quite a lot of time for calculation compared to other features, however according to (CVT, 2011) also much faster SIFT implementations exists (Vedaldi, 2006), and they should be used in real-time implementations.

3.5. Image-Based Control Algorithm

The relative satellite orientation according to the orbit frame (Wertz, 1978) is usually described by three Euler angles: roll ϕ_O, pitch θ_O and yaw ψ_O. If $\phi_O = \theta_O = \psi_O = 0$, the Earth observation point is defined by the intersection of Earth's surface and the line between the satellite and the center of the Earth.

To observe the desired point on the Earth's surface, the orbit frame is replaced by the reference frame, whose axis points to the desired point on the Earth's surface (see Figure 7).

Then the satellite body's orientation is described according to the reference frame, which is described by the three Euler angles: the roll ϕ, pitch θ and yaw ψ. When these angles approach zero, the satellite body frame aligns with the reference frame, and the satellite z axis (camera), therefore points to the desired point on the Earth. This is achieved by setting ϕ and θ. The angle ψ must also be set so that the observed image from the camera does not rotate but maintains some reference image rotation (the image contents do not rotate). The desired point on the Earth's surface and the reference frame orientation are, therefore, defined by the reference image's center and the rotation.

To meet the desired satellite orientation (attitude), the appropriate moments M (Eq. 11) around the satellite body frame axes are calculated, depending on the orientation errors. These orientation errors are estimated from the found feature pairs between the reference and the current image.

The error in the roll and pitch can be approximated from the estimated image translation (Δx , Δy), while the yaw is related to the rotation ($\Delta\varphi$) of the current image relative to the reference image.

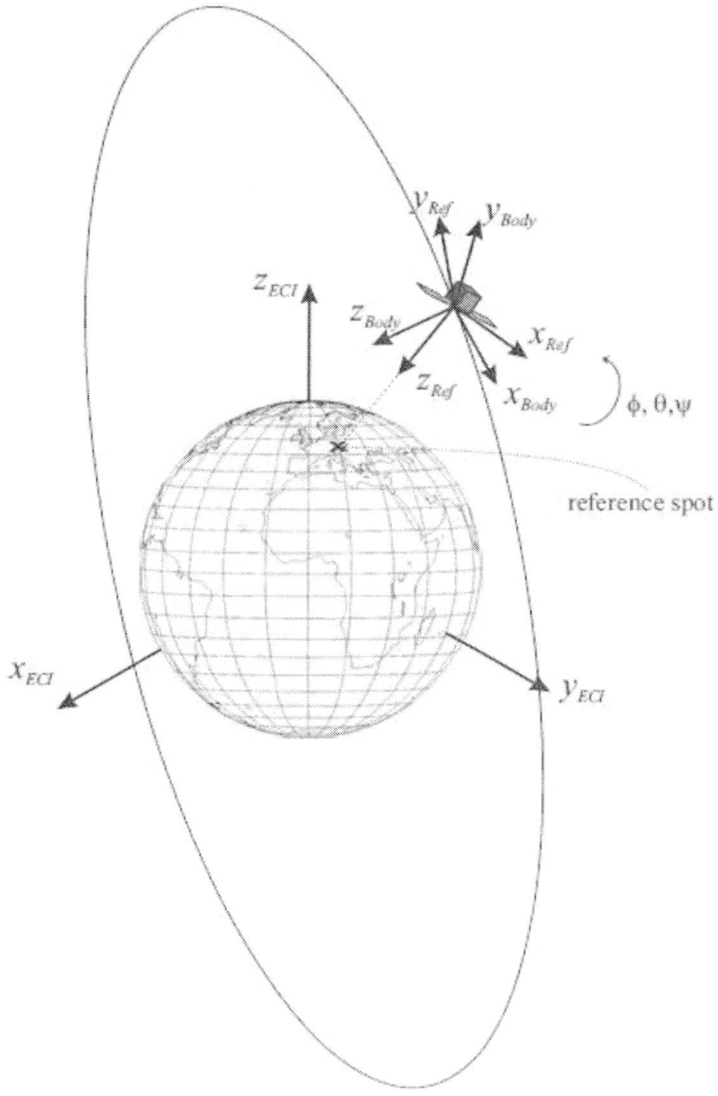

Figure 7. Illustration of ECI, reference and body coordinate systems.

The relation between the feature position in the reference image and the current image is as follows:

$$\begin{aligned} x_r &= \Delta x + x_c \cos\Delta\varphi - y_c \sin\Delta\varphi \\ y_r &= \Delta y + x_c \sin\Delta\varphi + y_c \cos\Delta\varphi \end{aligned} \tag{18}$$

where the indexes r and c denote the reference and the current image, respectively.

If the appropriate attitude control is applied to the satellite, then the image observed from the satellite camera nearly aligns with the reference image. Of course, some orientation error $\Delta\varphi$ is to be expected, mostly due to image-sensor noise, different camera perspectives, camera lens distortions and the like.

Nevertheless, the orientation error $\Delta\varphi$ is small, and therefore Eq. (18) can be simplified to

$$\begin{aligned} x_r &= \Delta x + x_c - y_c\Delta\varphi \\ y_r &= \Delta y + x_c\Delta\varphi + y_c \end{aligned} \tag{19}$$

By defining the displacement $d_x = x_r - x_c$ and $d_y = y_r - y_c$, the relation (19) in matrix form is $\mathbf{D} = \mathbf{P}T$ where $D_i = [d_{x\,i}, d_{y\,i}]^T$, $T = [\Delta x, \Delta y, \Delta\varphi]^T$ and $P_i = \begin{bmatrix} 1 & 0 & -y_{c\,i} \\ 0 & 1 & x_{c\,i} \end{bmatrix}$, $i = 1, \cdots, N$.

Where $P^T = \left[P_1^T \cdots P_N^T\right]$, $D^T = \left[D_1^T \cdots D_N^T\right]$ and N is the number of feature pairs found in the current and the reference images. The best estimate for the error vector T, considering all the feature pairs, is estimated using least squares:

$$T = (P^T P)^{-1} P^T D \tag{20}$$

The transfer-function relation between the input moments M and the orientation (see Eq. 11) is, by its nature, a double integral. Therefore, the PD control structure is used as follows:

$$M = \begin{bmatrix} k_{P_x}\Delta x + k_{D_x}\dfrac{d}{dt}(\Delta x) \\ k_{P_y}\Delta y + k_{D_y}\dfrac{d}{dt}(\Delta y) \\ k_{P\varphi}\Delta\varphi + k_{D\varphi}\dfrac{d}{dt}(\Delta\varphi) \end{bmatrix} \tag{21}$$

where k_{Px}, k_{Py}, $k_{P\varphi}$, k_{Dx}, k_{Dy} and $k_{D\varphi}$ are positive constants. Diagram of the proposed image-based attitude control algorithm is given in Figure 8.

By the proposed image-based approach, the calibration of intrinsic and extrinsic parameters of a camera is simple because the calibration is performed implicitly by tuning the controller parameters (21) ($k_{Px}, k_{Py}, k_{P\varphi}, k_{Dx}, k_{Dy}$, $k_{D\varphi}$), which results in the desired and stable operation. These controller parameters depend on intrinsic parameters (focal length of the camera and scale factors relating pixels to distance), but the latter are not required to be known and tuned explicitly. The calibration of extrinsic parameters (rotation and translation of the camera frame according to the main body frame) is not required because we control the view of the camera, which is also used as orientation sensor. In case of position-based control implemented by some other orientation sensor (e.g., star tracker), intrinsic parameters as well as extrinsic parameters (how satellite frame is related to the camera frame) also need to be known.

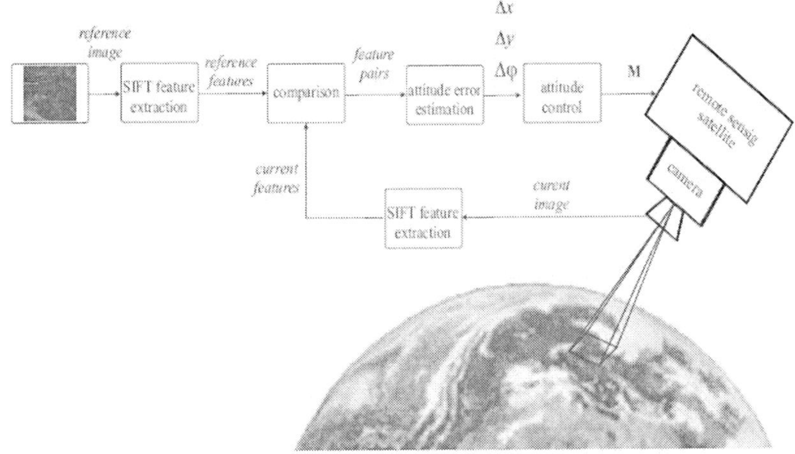

Figure 8. Image-based attitude control algorithm diagram.

Therefore, the proposed image-based approach enables much simpler calibration of the camera parameters compared to the position-based control implemented by some other external attitude sensor.

3.6. Classic Attitude Control

In this section, the attitude control law using exact satellite attitude information (which is hard to obtain in practice) is developed for comparison purposes.

The satellite attitude is compared to the reference satellite attitude; both attitudes are expressed by their Euler angles (roll, pitch and yaw) according to the ECI frame. The reference satellite frame is placed in the satellite center with the axis oriented to the desired Earth spot area (tourist place Bled in Slovenia, see Figure 7), while the axes are selected so that satellite camera image has the same orientation as the initial reference image.

In general, the rotation is defined by the rotation matrix (DCM) R_I^T, which transforms the orientation vector in the initial frame (v_I) to its representation in the target frame (v_T) and is defined by $v_T = R_I^T v_I$.

This rotation can also be represented by the Euler angles ϕ, θ and ψ (roll, pitch and yaw). First, the initial frame is rotated around the x axis for ϕ, then the newly obtained frame is rotated around its y axis for θ, and, finally, the newly obtained frame is rotated around its z axis for ψ. These Euler angles are obtained from the rotation matrix R_I^T as follows:

$$\begin{aligned} \phi &= \operatorname{arctan2}\left(\frac{R_I^T(3,2)}{R_I^T(3,3)}\right) \\ \theta &= -\arcsin\left(R_I^T(3,1)\right) \\ \psi &= \operatorname{arctan2}\left(\frac{R_I^T(2,1)}{R_I^T(1,1)}\right) \end{aligned} \tag{22}$$

where $\operatorname{arctan2}$ is the four-quadrant version of the inverse tangent function, and (i, j) are the indices in the rotation matrix R_I^T.

In our case, the rotation from the ECI to the satellite Body frame is defined by the R_{ECI}^{Body} rotation matrix (see Eq. 12) and by the Euler angles according to relation (22) ϕ_{Body}, θ_{Body} and ψ_{Body}. Similarly, the rotation from the ECI to the satellite reference frame is defined by the R_{ECI}^{Ref} rotation matrix and by the angles ϕ_{Ref}, θ_{Ref} and ψ_{Ref}. To determine the orientation-error angles between the satellite body frame orientation and the reference frame, the rotation from the body to the reference frame is defined by the rotation matrix $R_{Body}^{Ref} = R_{ECI}^{Ref} {R_{ECI}^{Body}}^T$. The error angles e_ϕ, e_θ and e_ψ are then calculated from R_{Body}^{Ref} (see Eq. 22).

The attitude control is then obtained like in (21), as follows:

$$M = \begin{bmatrix} k_{P\phi} e_\phi + k_{D\phi} \dfrac{d}{dt}(e_\phi) \\ k_{P\theta} e_\theta + k_{D\theta} \dfrac{d}{dt}(e_\theta) \\ k_{P\psi} e_\psi + k_{D\psi} \dfrac{d}{dt}(e_\psi) \end{bmatrix} \tag{23}$$

where $k_{P\phi}$, $k_{P\theta}$, $k_{P\psi}$, $k_{D\phi}$, $k_{D\theta}$ and $k_{D\psi}$ are positive constants.

4. Simulation Results

The proposed image-based control is verified using the satellite simulator in the Matlab environment. The position, orientation and focus point of the satellite are determined as presented in Section 2. To obtain realistic images of the Earth during the satellite camera motion, the Google Earth application is used to simulate the satellite's onboard camera. In the presented examples, the Kepler elements of the satellite orbit are taken from the Lapan Tubsat (Renner and Buhl, 2008) satellite, whose sun-synchronous orbit is approximately 600 km above the Earth's surface and needs approximately 90 minutes for one orbital period. The orbital speed of the satellite is approximately 7.5 km/s. The satellite's moment of inertia matrix is set to be the unit matrix.

The simulator was designed and selected for the presented experiments because it is more convenient to design and test the control-law operation in a simulation environment. The current position and orientation can be easily obtained in a simulator and used to verify the operation of the controller. However, in the case of a real satellite, the exact pose is hard to obtain, especially in the case of smaller satellites, which have a less precise attitude-determination system with fewer onboard sensors because of space limitations.

In the experiment, the satellite must constantly observe the desired spot on the Earth's surface. This spot (tourist place Bled, in our case) is defined by the initial reference image taken at time $t_0 = 25$ s, where the initial time 0 is the time of the Kepler elements definition. Prior to t_0, the satellite is moving in its orbit with zero angular velocities and no attitude control, in such an orientation that at $t_0 = 25$ s the camera focuses on the desired Earth observation spot. After t_0, the reference spot is defined, and the attitude control is switched on to enable a constant observation of that spot. The camera and attitude-control sampling frequency are fixed at 10 Hz and the camera resolution at 320 × 280 pixels.

The ground sampling resolution of the Google Earth image varies among different regions. The observed region in our case is covered by the resolution of 2.5m/pixel. The obtained images were down-sampled to 15m/pixel, which is comparable to real camera systems in the LEO (low Earth orbit) satellites where the resolution is typically 5-20m/pixel (Steyn, 2006).

In the following, the classic control is shown first, where the exact orientation data from the satellite motion simulator are used. Next the proposed image-based control operation is presented and evaluated.

4.1. Classic Control

The results of the simulation experiment using ideal attitude information are shown in Figure 9, where the satellite body attitude is compared to the reference satellite attitude, both with respect to the ECI frame. In the experiment, the satellite attitude is controlled using (23), where $k_{P\phi} = k_{P\theta} = k_{P\psi} = 0.1$ and $k_{D\phi} = k_{D\theta} = k_{D\psi} = 0.05$.

Figure 10 shows the error in the Euler angles, which approach zero after some initial transition phase. This means that the satellite camera (z axis) focuses on the required spot on the Earth's surface.

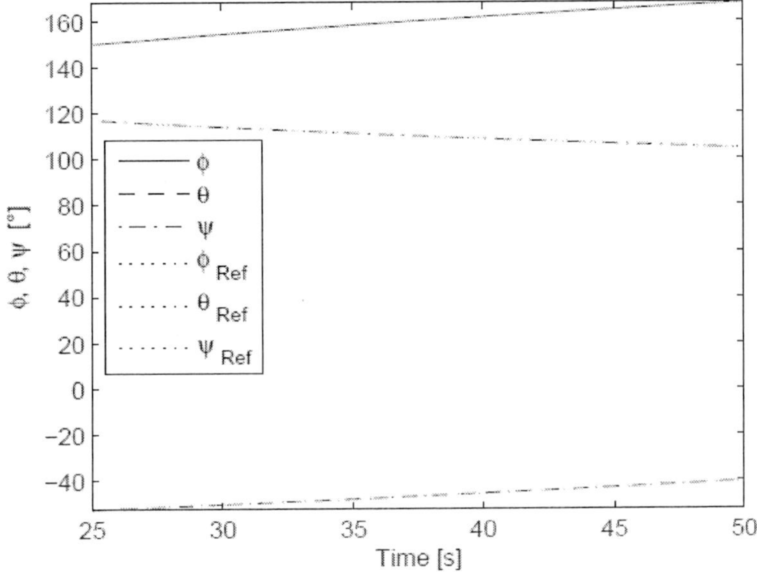

Figure 9. Satellite body frame and reference satellite body frame Euler angles with respect to ECI frame.

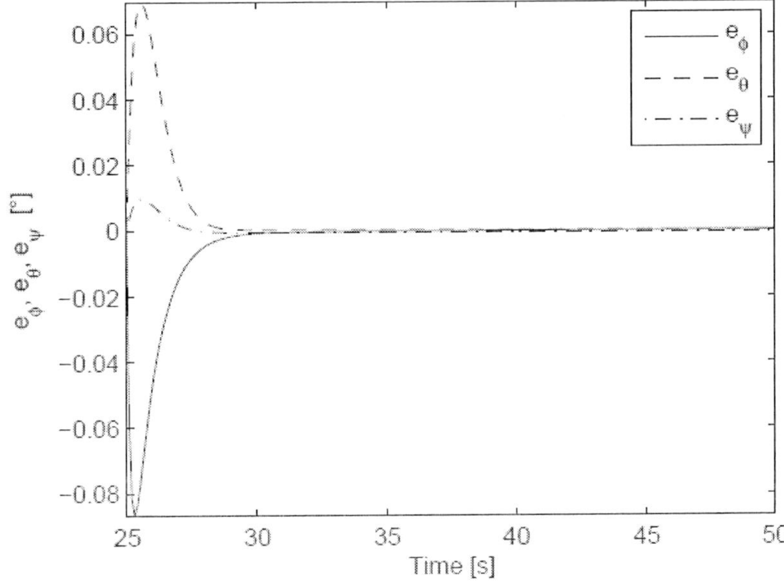

Figure 10. The error angles during the classic attitude control.

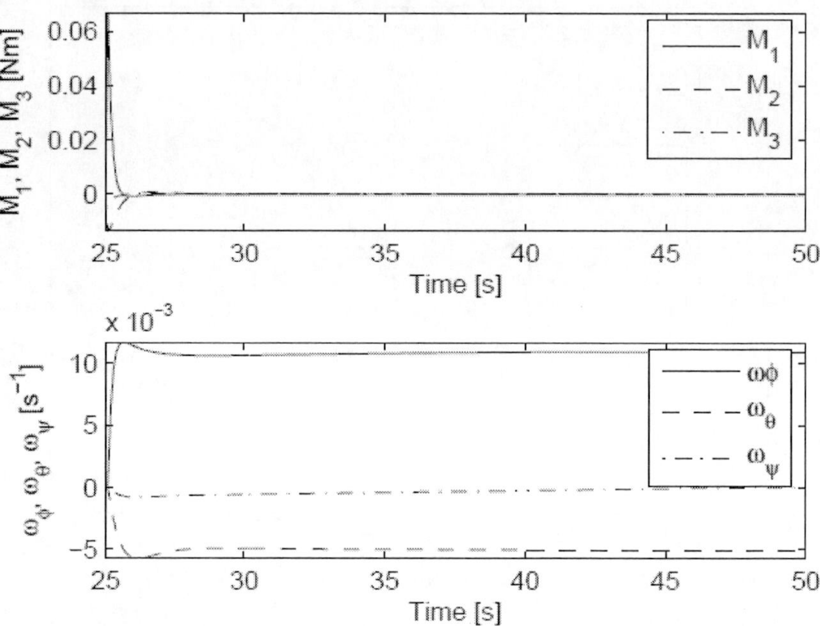

Figure 11. Moments and angular velocities applied to the satellite during the classic attitude control.

The initial transition phase appears because the satellite has zero angular velocities for all three axes at time t_0 when the attitude control starts. After the transition phase, when the satellite orientation errors approach zero, the appropriate satellite angular velocities are obtained, which enables constant tracking of the Earth reference spot.

The moments applied to the satellite during the control are shown in Figure 11. At the beginning of the experiment in Figures 10-11, some overshoot appears due to the initial transition phase.

The reference Euler angles in Figure 9 are approximately linear functions, with only a small contribution from the parabolic term and a negligible cubic-term contribution. The reference angles from Figure 9 can be approximated by polynomials of the third order, as follows:

$$\begin{aligned}\phi_{Ref}(t) &= 3.0\cdot 10^{-5}t^3 - 9.1\cdot 10^{-3}t^2 + 1.2t - 24.6\\ \theta_{Ref}(t) &= -4.2\cdot 10^{-5}t^3 + 6.5\cdot 10^{-3}t^2 + 0.21t - 8.7\\ \psi_{Ref}(t) &= -3.6\cdot 10^{-5}t^3 + 10.2\cdot 10^{-3}t^2 - 1.0t + 19.5\end{aligned} \quad (24)$$

which means that the proposed controller structure (23) is appropriate for following the reference without any tracking error or with only a small, constant tracking error due to the small parabolic part. The integral part could also be added to the control (23) to cancel the tracking error if the reference angles have a higher parabolic contribution.

4.2. Image-Based Attitude Control

As already mentioned, our goal is to control the satellite attitude using image sensor information only. In this section, the simulation results using the proposed image-based attitude control (IB control) are presented. The proposed IB control (Eq. 21) is used with $k_{Px} = k_{Py} = 2.5 \cdot 10^{-4}$, $k_{P\varphi} = 1.4$, $k_{Dx} = k_{Dy} = 1.5 \cdot 10^{-4}$ and $k_{D\varphi} = 1.8 \cdot 10^{-1}$.

The satellite path projection on the Earth's surface and the trajectory of the camera focus point for an uncontrolled satellite are shown in Figure 12. Because no control is applied, the satellite does not correct its orientation, and therefore the camera focus travels over the Earth's surface.

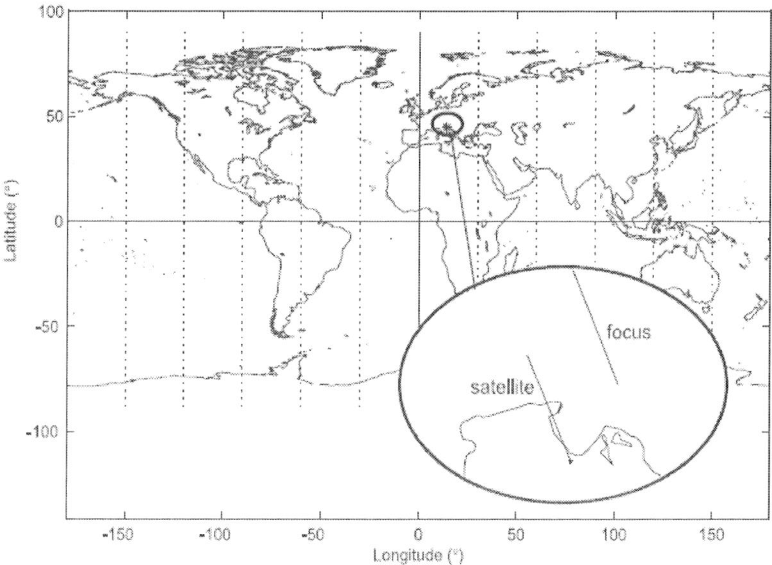

Figure 12. Satellite motion and camera focus point without image-based control.

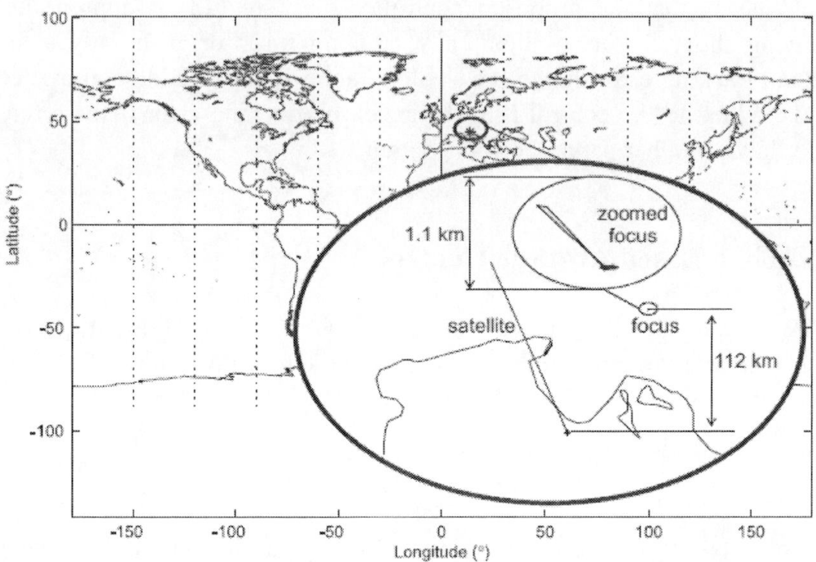

Figure 13. Satellite motion and camera focus point with image-based control.

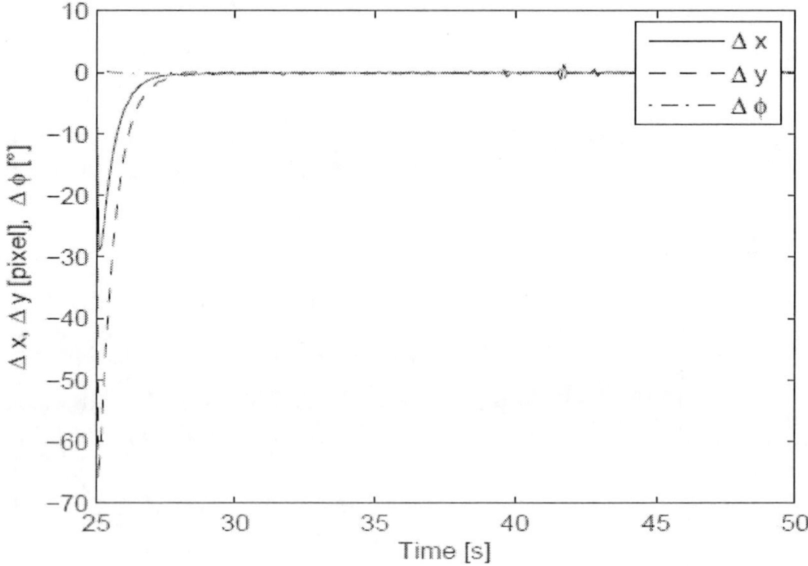

Figure 14. Satellite IB attitude control. Tracking errors (translations Δx , Δy and rotation $\Delta\varphi$) between the reference image and the current image.

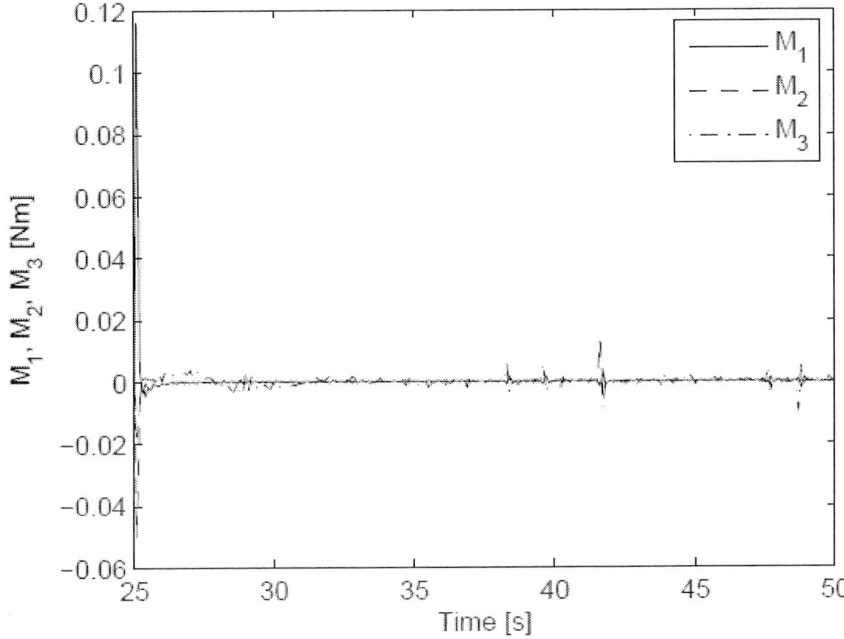

Figure 15. Moments applied to the satellite during IB attitude control.

In Figure 13, the experiment with image-based control is shown. In this experiment, the camera is focused on a fixed point on the Earth's surface that corresponds to the tourist place Bled in Slovenia. At the beginning of the experiment, some overshoot appears. After this, the camera focus point is quite stable.

A small amount of noise is observed in the zoomed part of Figure 13, which is due to the SIFT feature calculation and a least-squares estimation of the image translation and rotation.

The results of the image-based control are shown in Figure 14, where the tracking errors Δx, Δy, $\Delta \varphi$ are indicated. These errors are obtained from comparisons of the current and the reference images using least-squares estimation (20). Some control-error oscillations are observed, which is due to the SIFT features' estimation and because the current images are observed with a different perspective, rotation and scale than the reference image.

The moments applied to the satellite during control are shown in Figure 15. As already mentioned, the SIFT features are quite robust to translation, rotation and scale in some reasonable area.

However, if this area is exceeded (e.g., for rotations of $\pm 15^{\circ}$), the estimated SIFT features are less reliable, which causes an increase in the attitude control error. Additionally, if the same scene as in the reference image is observed at a different perspective, the estimated image frame axes are not exactly the same as in the reference image.

One of the reasons for this is perspective, which causes the reference axes (x and y) to be no longer perpendicular. The other reason could be the height of the reference Earth spot, which is not considered. The reference spot when observed from the side angle appears to be at a different position than it really is.

The control tracking errors (Figure 14) used in the IB control loop quickly approach zero, which confirms the correct and stable operation of the proposed control system.

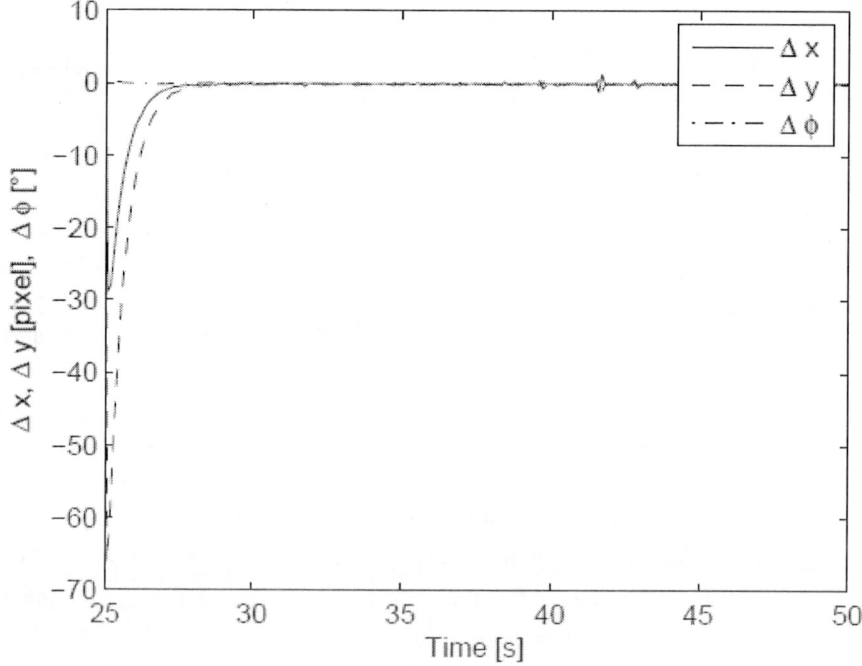

Figure 16. Satellite IB attitude control with moving reference. Tracking errors (translations Δx, Δy and rotation $\Delta\varphi$) between the moving reference image and the current image.

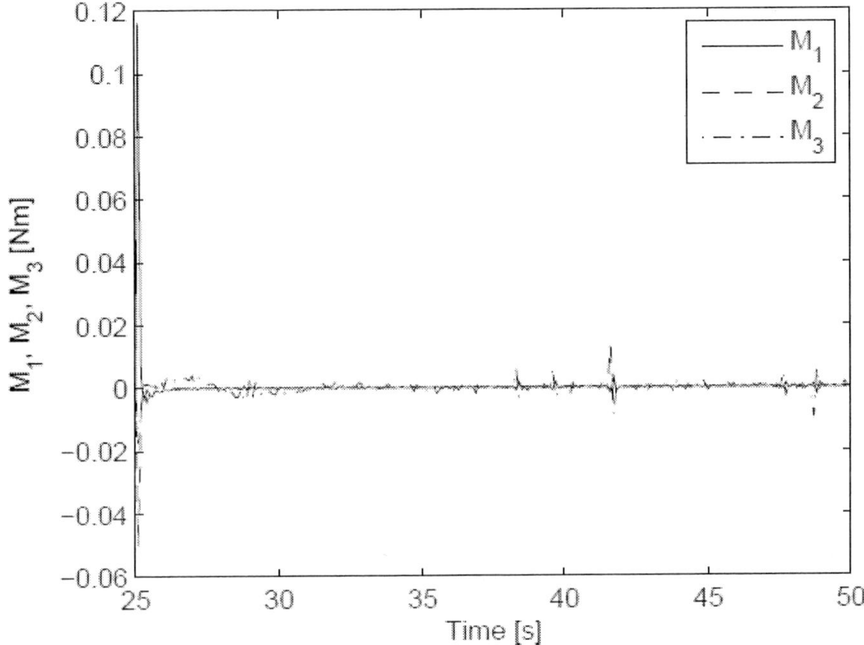

Figure 17. Moments applied to the satellite during IB attitude control with moving reference.

A practical solution that increases the proposed IB control system's robustness to the SIFT features' estimation, perspective and some other unmodelled effects is the use of a moving reference. In the next experiment, in Figures 16–17, the reference image is changed to the current image every $T_{ref} = 4$ s.

The moments applied to the satellite during the control are shown in Figure 16.

Compared to the previous experiment (Figures 14-15), the changes of rotation, scale, perspective and other influences have less effect on the attitude control error, while the quality of the tracking for the image centre and rotation do not change significantly. Also, the oscillations of the signals in steady state (Figure 16 and Figure 17) are lower, and the time of the experiment can be much longer.

In the previous experiment, the satellite can focus on the reference spot until t=158 s, after which the control fails because not enough reliable feature pairs can be found. In the case of the moving reference, the control can reliably operate until the camera image has some useful content or until the camera view to the reference spot becomes blocked by the Earth surface. The tracking errors are similar as in the fixed reference approach, but the noise level is decreased and the robustness, operational time and approach applicability are increased. Because of many features pairs found (cca. 350 pairs) in the image, the control tracking error is averaged and is sub pixel. According to Figure 16, the control tracking error is 0.2 pixels, which correspond to $3\cdot10^{-4}$ degrees or 3.1m on the Earth surface. This means that obtained orientation of the camera view is quite stable with the image jitter less than 0.2 pixels.

To verify the operation of the proposed IB moving-reference control approach, the comparison between the obtained satellite images and the ideal images is shown in Figure 18. Ideal images would be obtained if the satellite focuses perfectly on the desired Earth spot. The desired Earth spot is defined by the center of the first image taken at time $t_0 = 25$ s.

Note that the goal of IB control is to force the translational and rotational errors (obtained from Eq. 20) to zero. These average errors result from all the feature pairs found in the reference image and the current image. This is why it is not possible to define the error in the rotation for the whole image. Due to the change in the perspective, different angles from the reference image are distorted in a very different way for the consequent images. In the IB control, the rotation depends heavily on the direction containing more features. Nevertheless, the image content obtained from the camera is still very close to the ideal images, as seen in Figure 18.

The satellite tracking errors ε_x and ε_y (according to the first image at $t = t_0$) at $t = 50$ s are approximately three pixels, which corresponds to a 47 m translation error that is hardly noticeable on the camera image (see Figure 18). Actual satellite attitude is compared also to the reference satellite attitude both with respect to the ECI frame (Figure 19). Reference attitude Euler angles is determined so that satellite z axis always points to the desired Earth observation point (roll an pitch) and rotation around z axis (yaw) so that y satellite axis points in the direction of north. Actual satellite attitude is available in simulator, but in real situations only its estimate can be obtained if satellite is equipped with appropriate attitude localization unit.

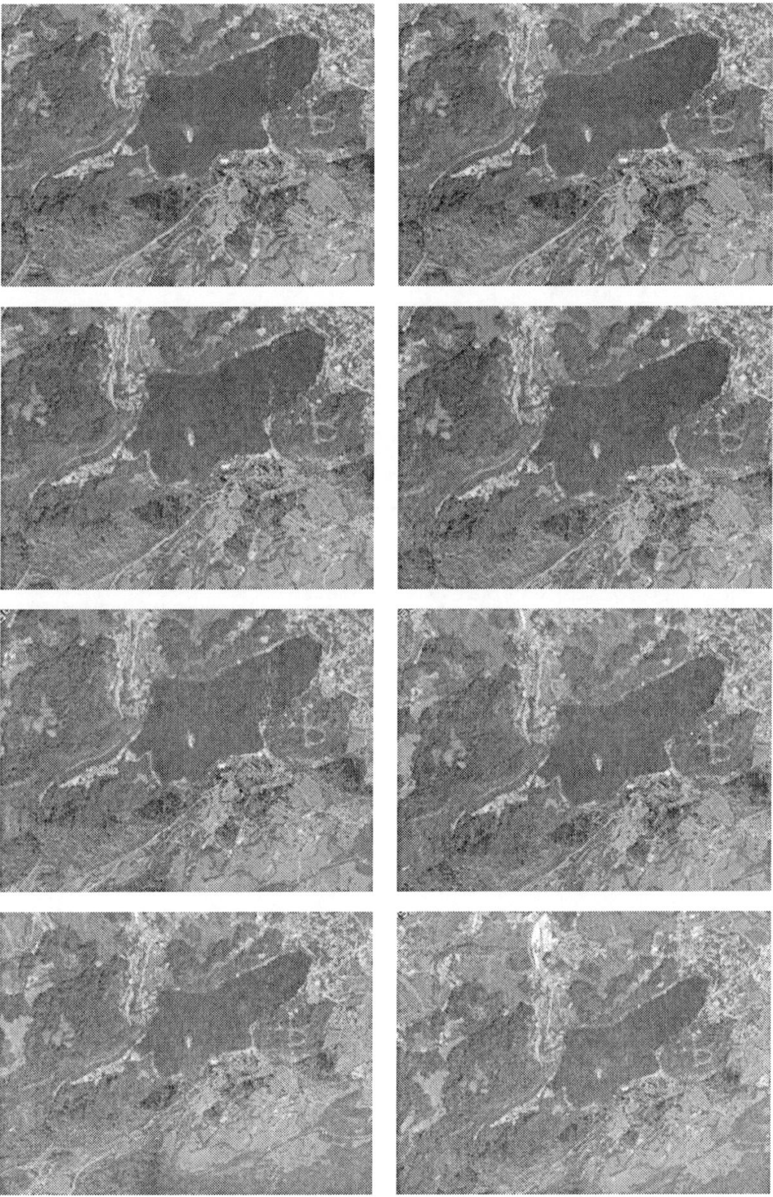

Figure 18. Sequence of images during satellite motion. In the first column, the images obtained with IB control and the moving reference are shown; they are taken at: 25 s, 50 s, 75 s and 100 s. In the second column, the ideal images that would be obtained if the satellite perfectly tracks the desired Earth observation spot are shown; they are taken at: 25 s, 50 s, 75 s and 100 s.

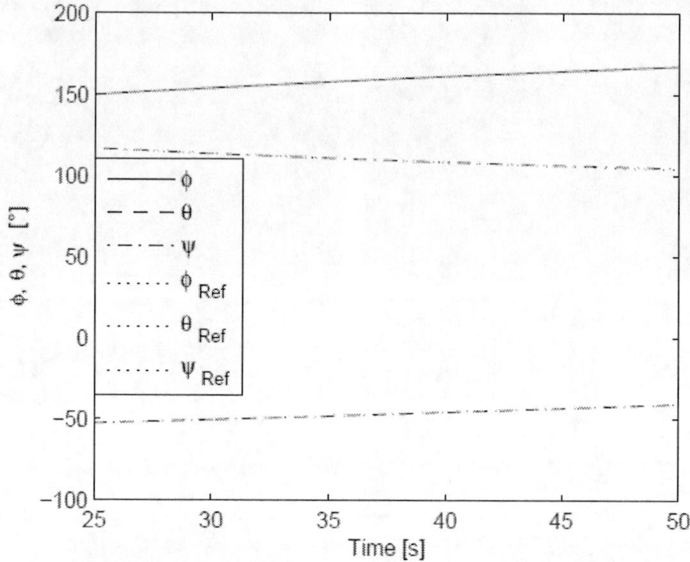

Figure 19. Reference and body Euler angles according to ECI frame during IB control. Euler angles are calculated from the exact satellite attitude obtained from the simulator.

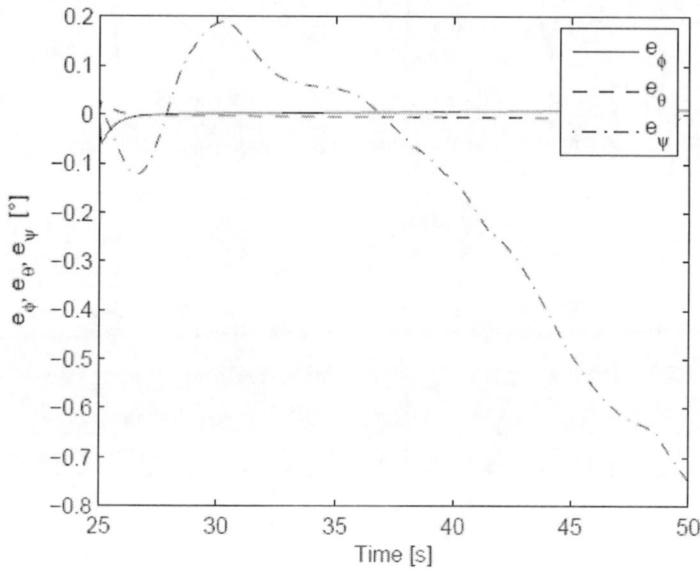

Figure 20. Error Euler angles during IB control. Errors are calculated from the exact satellite attitude obtained from the simulator.

Table 1. IB control tracking performance comparison for the fixed ($_f$) and moving ($_m$) references

	30 s	50 s	100 s	150 s	200 s	250 s	280 s
$\varepsilon_{x\,f}$ [pixel]	1	1	3	32	–	–	–
$\varepsilon_{y\,f}$ [pixel]	0	1	7	16	–	–	–
$\sigma_{x\,f}$ [pixel]	0.07	0.42	1.43	21.34	–	–	–
$\sigma_{y\,f}$ [pixel]	0.10	0.62	1.62	12.87	–	–	–
FP_f	324	296	110	18	–	–	–
$\varepsilon_{x\,m}$ [pixel]	1	-1	1	5	13	29	27
$\varepsilon_{y\,m}$ [pixel]	1	2	3	9	25	-3	-15
$\sigma_{x\,f}$ [pixel]	0.04	0.23	0.21	0.28	0.21	3.23	30.42
$\sigma_{y\,f}$ [pixel]	0.03	0.19	0.09	0.12	0.08	2.31	14.1
FP_m	335	330	290	285	225	143	24

The real satellite attitude error is shown in Figure 20, where error in roll and pitch at $t = 50$ s is approximately 0.005°, which at 600 km height correspond to 47 m translation error, while the maximal error in yaw is 0.75°, which is hardly noticeable on the camera image.

These conclusions can also be confirmed from Table 1, where the tracking errors ε_x, ε_y and their standard deviations (σ) for a fixed image reference (subscript $_f$) and for a moving image reference (subscript $_m$) are shown. Note that the rotation error is not shown, due to the fact that it cannot be defined uniquely, as explained before. The tracking errors in Table 1 are evaluated according to the first image center at $t = t_0$. As already explained, the image distortions due to perspective and error due to the lower SIFT feature reliability at higher rotation angles are not considered by transformation model (19). Therefore, the center of the current image has some additional error according to the reference image center. The influence of these errors to the tracking accuracy of the current image center is analyzed in Table 1. At longer observation times, the reference spot is observed by a higher tilt angle, and the image becomes distorted because of a change in perspective. At times, $t > 280$ s the reference spot cannot be seen because the satellite view becomes blocked by the Earth (the hills around Bled cover the reference spot). Nevertheless, the estimated tracking error of the camera image center is approximately three pixels (for $t < 150$ s and moving reference approach), which corresponds to 47 m translation error. Note that this additional error does not make the proposed IB control unstable, because it is not included in the control tracking error (Figure 16). The approach with a moving reference can follow the desired Earth spot for a longer time (until $t = 280$ s, see Table 1). The number of found feature pairs (FP) between the reference image and the current image for the moving-reference approach is higher and more constant, which also explains the lower attitude-control error oscillations. A movie clip of the experiment can be seen at http://msc.fe.uni-lj.si/PublicWWW/Klancar/IBCmovie.html.

4.3. Performance Evaluation of the IB Control Strategy

Here, the performance of the proposed image-based attitude control and position-based attitude control with external sensor are analyzed. The obtained tracking results are compared to the expected accuracy of sun sensors and star tracker sensors that are usually used for attitude determination (Steyn, 2006; Wertz, 1999). The true orientation of the satellite (obtained from the simulator) was corrupted by the normal noise with standard deviation set according to the sun sensor expected accuracy (0.1°). To compare control tracking results, the same feedback and PD controller gains are used as in the case of the proposed image-based control. The simulated sun sensor measurements, therefore, need to be scaled from radians to the pixels. This procedure enables us to figure out how the external sensor noise propagates to the tracking error in picture coordinates. The results of the attitude control performance analysis for sun sensor and star tracker external sensors are compared to the proposed image-based attitude control. The results are summarized in Table 2. The attitude control using sun sensor (usually combined with magnetometer) results in 90 pixels tracking error, which corresponds to 1340 m ground tracking error. Attitude control using star tracker sensor results in one pixel tracking error, which corresponds to 15 m ground tracking error. As already evaluated, the proposed image-based attitude control has 0.2 pixels tracking error noise, which corresponds to 3.1 m ground tracking error noise. All these errors correspond to a 95% confidence interval. The estimated error of tracking the reference image center includes also the systematic error due to perspective and is three pixels or 47 m on the ground. According to the analysis in Table 2, the expected accuracy (where both tracking and systematic errors are considered) of the image-based attitude control for tracking the reference image center is approximately 28 times higher and three times lower than the accuracy of sun sensor and star tracker, respectively.

Table 2. Performance evaluation of attitude control using external sensor (sun or star tracker sensor) and the proposed image-based (IB) strategy

sensor	sensor accuracy [°]	control tracking error [pixel]	control tracking error [m]
sun sensor	0.1	90	1340
star tracker	0.001	1	15
IBC - feature tracking	–	0.2	3.1
IBC - image center tracking	–	3	47

The proposed approach is, therefore, close to the star tracker sensor performance, which is expensive, big and heavy (1 kg or more) and is therefore less appropriate for small LEO observation satellites.

CONCLUSION

Image-based approach is presented to control the remote sensing satellite so that desired reference point on the Earth's surface is observed. The approach is derived and demonstrated in a simulation environment. To simulate the satellite motion in the orbit, the satellite kinematics and dynamics are modelled. The camera image is simulated using the Google Earth application, and SIFT features are used to implement image-based control.

According to the performed analysis of the usually used image feature detectors, the SIFT features are appropriate for such a task because they are invariant to image translation, scaling, rotation, and partially invariant to illumination changes and an affine or 3D projection.

They are, however, computationally demanding, which consequentially lowers the response time of the real-time control. A lower image resolution must be used; however, this increases the tracking error of the control. Future work idea is optimization of SIFT feature calculation or using some alternative feature generator such as SURF.

To improve the robustness of the proposed control to measurement noise, perspective, and other unmodelled effects, the concept of a moving reference is implemented. The obtained tracking performance is similar to the case with the fixed reference, while the performance regarding noise, robustness and operational time is greatly improved. The presented image-based control showed good performance in simulations with realistic assumptions, so it could also be applicable to real satellite platforms. Performance analysis results show that the proposed approach accuracy is close to the performance of the star tracker sensor and much better than the expected accuracy of the sun sensor.

Acknowledgments

The presented work has been performed within the Centre of Excellence for Space Sciences and Technologies SPACE-SI which is an operation partly financed by the European Union, European Regional Development Fund and Republic of Slovenia, Ministry of Education, Science, Culture and Sport. The work was also supported by research program P2-219 "Modelling, simulation and control of processes" . The authors would also like to thank Andrej Zdešar for providing Matlab code for image feature detectors used in the performance analysis.

References

Bai, H.; Wan, E.; Song, X.; Myronenko, A. (2007). Vision-only Navigation and Control of Unmanned Aerial Vehicles Using the Sigma-Point Kalman Filter, Proceedings of the 2007 National Technical Meeting of The Institute of Navigation; San Diego, CA, pp. 1264-1275.

Cesetti, A.; Frontoni, E.; Mancini, A.; Zingaretti, P.; Longhi, S. (2010). A Vision-Based Guidance System for UAV Navigation and Safe Landing using Natural Landmarks, *Journal of Intelligent and Robotic Systems*. DOI 10.1007/s10846-009-9373-3, Vol. 57, No. 1-4, pp. 233-257.

Erhard, S.; Wenzel, K.E.; Zell, A. (2010). Flyphone: Visual Self-Localisation Using a Mobile Phone as Onboard Image Processor on a Quadrocopter, *Journal of Intelligent and Robotic Systems*. DOI 10.1007/s10846-009-9360-8, Vol. 57, pp. 451–465.

Fan, Y.; Ding, M.; Liu, Z.; Wang, D. (2007). Novel remote sensing image registration method based on an improved SIFT descriptor, in: *Society of Photo-Optical Instrumentation Engineers (SPIE) Conference Series*; Vol. 6790, doi: 10.1117/12.751479, pp. 67903G.1-67903G.6.

Hoots, F.R.; Roehrich, R.L. (1988). Spacetrack Report No.3, *Department of Commerce National Technical Information Service*, Springfield, VA.

Klančar, G.; Blažič, S.; Matko, D.; Mušič, G. (2011). Image-Based Attitude Control of a Remote Sensing Satellite, *Journal of Intelligent and Robotic Systems*. doi: 10.1007/s10846-011-9621-1.

Lowe, D.G. (2004). Distinctive image features from scale-invariant keypoints, *International Journal of Computer Vision*. Vol. 60, Vo. 2 , pp. 91-110.

Lowe, D. (2005). Demo Software: SIFT Keypoint Detector. Available at: http://www.cs.ubc.ca/ lowe/keypoints/.

Psiaki, M.L.; Martel, F.; Pal, P.K. (1990). Three-Axis Attitude Determination via Kalman Filtering of Magnetometer Data, *Journal of Guidance, Control, and Dynamics*. Vol. 13, No. 3, pp. 506-514.

Raj, E.S.; Venkatraman, S.; Varadan, G. A. (2008). Fuzzy approach to Region-of-interest coding in JPEG-2000 for ATR Applications from High Resolution Satellite Images, in ICVGIP '08: Sixth Indian Conference on Computer Vision, Graphics and Image Processing, *IEEE Computer Society*; pp. 1993-200, doi: 10.1109/ICVGIP.2008.53.

Renner, U.; Buhl, M. (2008). High precision interactive Earth observation with Lapan-Tubsat, in Proceedings of the 4S Symposium Small Satellites, Systems and Services, Rhodos, Greece, pp. 26-30.

Se, S.; Little, D.L.J. (2002). Mobile robot localisation and mapping with uncertainty using scale-invariant visual landmarks, The International *Journal of Robotics Research*. Vol. 21, No. 8, pp. 735-758.

Steyn, W.H. (2006). A view finder control system for an Earth observation satellite, *Aerospace Science and Technology*. Vol. 10, pp. 248–255.

Škrjanc, I. (2006). Pitch angle control of unmanned air vehicle with uncertain system parameters, *Journal of Intelligent and Robotic Systems*. Vol. 47, No. 3, pp. 285-297.

Tuytelaars, T.; Mikolajczyk, K. Local Invariant Feature Detectors: *A Survey, Computer Graphics and Vision*. 2007, Vol. 3, No. 3, pp. 177–280.

Vedaldi, A. (2006). SIFT++. Available at: http://www.vlfeat.org/ vedaldi/code/siftpp.html.

Wertz, J.R. Spacecraft Attitude Determination and Control; D. Reidel Publishing Company: Dordrecht, Holland, 1978.

Wicks, A.; Silva-Curiel, A.; Ward, J.; Fouquet, M. (2000). Advancing Small Satellite Earth Observation: Operational Spacecraft, Planned Missions And Future Concepts, in Proceedings of 14th Annual AIAA/USU Conference on Small Satellites; Logan, UT, pp. 1-8.

Wertz, J. R.; Larson, W. J. (1999). Space mission analysis and design; Kluwer academic publishers.

Wong, A.; Clausi, D.A., (2009). AISIR: Automated inter-sensor/inter-band satellite image registration using robust complex wavelet feature representations, Pattern Recognition Letters. doi: 10.1016/j.patrec.2009.05.016.

Zdešar, A. (2012). Matlab toolbox for image feature detection. Available at: http://msc.fe.uni-lj.si/PublicWWW/andrej/libs/index.html.

CVT-Computer vision talks (2011). Comparison of the OpenCV's feature detection algorithms. Available at: http://computer-vision-talks.com/2011/01/comparison-of-the-opencvs-feature-detection-algorithms-2/.

In: Remote Sensing
Editor: Enner Alcântara

ISBN: 978-1-62417-140-6

Chapter 6

Assessing Urban Growth and Soil Sealing with Spectral Mixture Analysis

Ignacio Melendez-Pastor*, Jose Navarro-Pedreño, Ignacio Gómez and Encarni I. Hernández
Department of Agrochemistry and Environment
University Miguel Hernández of Elche
Elche (Alicante), Spain

Abstract

The world population is in continuous growth, more pronounced in the population living in urban areas than those in rural spaces. This increase in urban population results in a massive growth of cities in order to accommodate the people and their activities. A result of urban growth is the loss and sealing of agricultural and natural soils. This loss of productive soils has negative consequences for food production and natural soil functions as carbon sequestration. In recent decades, remote sensing has played an important role in monitoring urban growth. There are numerous applications where, through the use of aerial or satellite imagery, researchers have been able to quantify the temporal evolution of the surface occupied by urban areas or the percentage of impervious surfaces of the cities. However, there are great difficulties using medium

* E-mail: imelpas@gmail.com

spatial resolution remote sensing imagery (i.e., Landsat-type) for mapping urban areas due to the mixture of covers (e.g., asphalt, concrete, gardens) of the cities. In addition, the boundaries of under-construction urban areas or scattered residential areas are fuzzy and commonly confused with natural or agricultural areas. This study aims to evaluate the process of urban growth in the south province of Alicante from 1999 to 2007. A methodological approach combining unsupervised/supervised classification and spectral mixture analysis was employed to obtain land cover maps from satellite remote sensing data. This test area combines relevant factors to evaluate the procedure, such as a significant population growth and urban expansion, the presence of large gardening areas mixing urban spaces that produce a mixture of covers, and the coexistence of dense residential/industrial areas and scattered residential areas in an agricultural environment. An initial ISODATA unsupervised classification was employed to discriminate pure clusters uniquely labeled with a land cover class from mixed clusters masked for subsequent analyses. Spectral mixture analysis was applied to those mixed class areas. Thus, different combinations of endmembers were tested to obtain precise spectral mixture models to characterize land cover mixed areas.

A classification and regression tree (CART) algorithm employed the endmember fraction images to map land covers in the mixed class areas. Good overall agreement and Kappa coefficient (~ 0.9) values were reported. Major land cover change was the increase of urban areas (+ 127%) diminishing shrub and orchard areas. Huge soil sealing processes promoted a very negative impact on the soil resource.

1. Introduction

Historically, increasing urban population has driven the growth of cities. However, even where there is little or no population pressure, a variety of factors are still driving sprawl in many areas. These are rooted in the desire to realize new lifestyles in suburban environments, outside the cities (EEA, 2006a). In addition, there are also phenomena of demand for housing as a second residence in tourism areas without involving increases in the population census. Both processes are often characterized by the demand for residential houses. In this sense, the term urban sprawl has been commonly used to describe this physical pattern of low-density expansion in large urban areas, under little planning control of land subdivision, mainly into the surrounding agricultural areas (EEA, 2006a).

The problem is that urban sprawl has severe impacts on natural resources (loss of agricultural land, soil sealing, landscape fragmentation) and implies many socio-economic changes (social polarization, economic inefficiency) in these areas. The construction of new low-density residential areas also requires new communication infrastructure that increases the consumption of land. This implies a massive soil sealing to build houses or pave roads.

Soil sealing is the destruction or covering of soils by buildings, constructions and layers of completely or partly impermeable artificial material (i.e., asphalt, concrete, etc.), being the most intense form of land-take, and is essentially an irreversible process (Prokop et al., 2011). When land is sealed, soil functions are reduced and may have a great impact on surrounding soils by changing water flow patterns and increasing the fragmentation of biodiversity (European Commission, 2002).

Soil sealing interrupts the exchange in between the soil system and other ecological compartments, leading to a number of negative effects on the soil system such as (Prokop et al., 2011): (1) less availability of fertile soils for future generations; (2) reduction of soil functions; (3) loss of water retention areas and increase in surface water runoff; (4) less soil carbon sequestration and carbon storage; (5) landscape fragmentation and loss of biodiversity through reduction of habitats and remaining systems too small or too isolated to support species; (6) unsustainable living patterns by urban sprawl; and (7) sealed surfaces have higher surface temperatures than green surfaces and alter the microclimate in particular in highly sealed urban areas.

In order to understand the impact of urban growth in natural resources like soil, it is necessary to map the evolution over time of urban areas and their natural surroundings. Geographical information systems (GIS) and remote sensing (RS) techniques introduced a new era for land resources assessment and monitoring in terms of information quality (Mermut and Eswaran, 2001). RS and GIS technologies are highly compatible primarily because of the nature of RS as a source of spatial land use/land cover information to be merged with other datasets in GIS for environmental applications (Nizeyimana, 2006).

However, urban areas are composed of a diverse assemblage of materials arranged by humans in a complex way, which make the spectral recognition of urban landscapes greatly difficult (Jensen, 2007). So, remote sensing of spatial and temporal changes in urban areas face the following challenges (Small, 2002): (1) to accurately determine the relative contribution of different endmembers to individual reflectance measurements; and (2) to compensate for temporal variations in observed radiance that are not related to changes in surface reflectance. To deal with such challenges, many researchers have investigated the application of spectral unmixing to map and understand the spatial and temporal dynamics of urban sprawl areas (Alberti et al., 2004; Lu and Weng, 2004; Weng et al., 2006; Michisita et al., 2012).

The objective of this chapter was the multi-temporal analysis of urban growth in a coastal tourism area of southeast Spain. In our approach, spectral unmixing and hard land cover classification techniques (supervised and unsupervised) were combined to obtain land cover maps of sufficient quality to ensure a proper multi-temporal analysis.

2. Material and Methods

The study area is the south of the province of Alicante (Southeast Spain) bounded in the north by the Segura River.

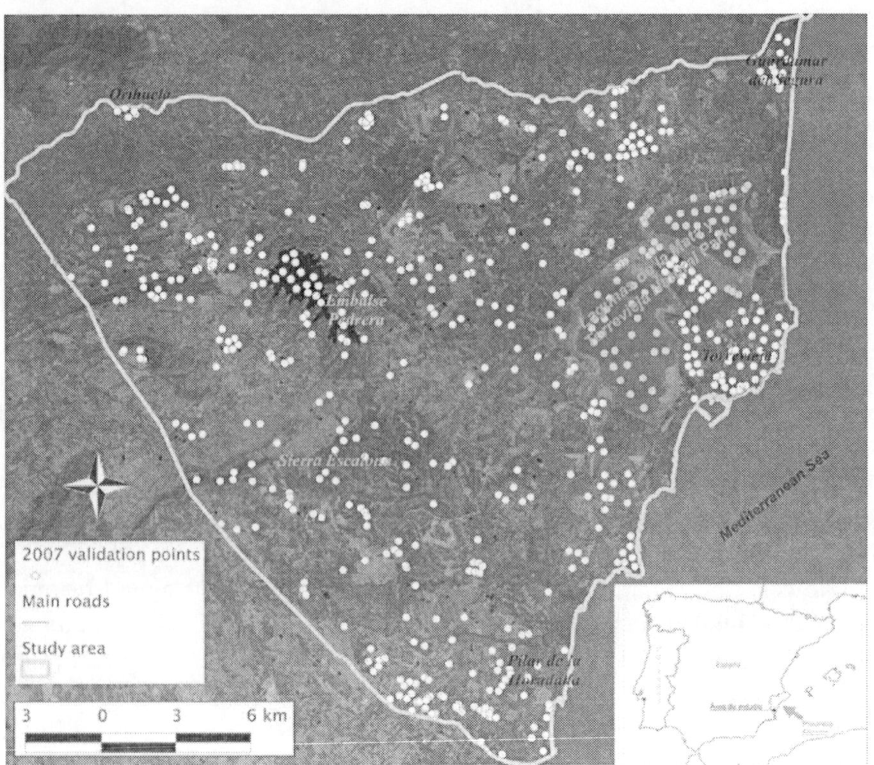

Figure 1. Map of the study area. Land cover validation points for 2007 are shown.

The area covers 539 km^2 (Figure 1) and comprises Tertiary coastal hill as part of the Betic Mountains and Quaternary alluvial plains resulting from the accumulation of sediments from the Segura river and temporary watercourses. Nowadays, this is a coastal urban area, with scattered residential houses, irrigated crops, Mediterranean coniferous forest (mainly in the Sierra de Escalona) a large reservoir (Pedrera reservoir) and two salt lakes as part of the Lagunas de La Mata y Torrevieja Natural Park and included in the RAMSAR list of wetlands of international importance.

Torrevieja is the most populated municipality in the study area (106,000 permanent inhabitants in 2012) and an important European tourism destination. The climate is semiarid Mediterranean with a mean annual rainfall about 300 mm and a mean annual temperature of 17 °C. The summer season is hot and dry while winter season is warm.

2.1. Satellite Imagery and Preprocessing Methods

Remote sensing data were acquired by the Thematic Mapper (TM) and Enhanced Thematic Mapper Plus (ETM+) sensors on-board the Landsat 5 and Landsat 7 satellites, respectively. A scene acquired on 21st July 1999, by the ETM+ sensor and a scene acquired on 4th August 2007, by the TM sensor were employed for analysis. Both scenes were recorder at the path 199 and row 33 of the Worldwide Reference System (WRS) 2 employed by Landsat satellites. Typically, summer meteorological conditions without cloud coverage were registered on the date of image acquisition, and thus the image quality was optimal. Date of image acquisition was similar (mid-summer), ensuring multi-temporal comparability.

Satellite image preprocessing included geometric and atmospheric corrections to ensure the spatial and temporal comparability and to obtain at-ground reflectance pixel spectra. Digital vector cartography (1:10,000) and aerial orthophotos (0.5 m of pixel resolution) were used for the geometric correction. The Landsat 7 ETM+ scene was the base image for the geometrical correction by its better visual quality. Subsets of the Landsat scenes covering the study area were used for analysis. The Landsat ETM+ image was geometrically corrected using Ground Control Points (GCP) identified on the orthophotos and cartographic maps.

A quadratic mapping function of polynomial fit and the nearest-neighbor resampling method were used for the correction. The Landsat TM image was co-registered to the ETM+ base image employing additional GCPs. The nearest-neighbor resampling method was selected because it ensures that the original (raw) pixel values are retained in the resulting output image, which is an important requirement in any change detection analysis (Mather, 2004). The maximum allowable root mean square error (RMSE) of the geometric correction was less than half a pixel, a reference value frequently cited (Alberti et al., 2004; Jensen, 2005). Atmospheric correction involves the estimation of the atmospheric optical characteristics at the time of image acquisition before applying the correction to the data (Kaufman, 1989). This type of correction is a pre-requisite in many remote sensing applications such as in classification and change detection procedures (Song et al., 2001). Radiometric calibration was applied prior to the atmospheric correction. The conversion of raw digital numbers (DN_{raw}) of Landsat level 1 (L1) image products to at-satellite radiance values (L_{sat}) required the application of current re-scaling values (Chander et al., 2010) by applying the following expression (Chander and Markham, 2003; Irish, 2012; Chander et al., 2010):

$$L_{sat} = \left(\frac{L_{MAX\lambda} - L_{MIN\lambda}}{255} \right)(DN) + L_{MIN\lambda} \tag{1}$$

where L_{sat} is at-satellite radiance [W/(m^2 sr μm)]; $L_{MIN\lambda}$ is the spectral radiance that is scaled to Q_{calmin} [W/(m^2 sr μm)] (Q_{calmin} is the minimum quantized calibrated pixel value, i.e. DN=0, corresponding to $L_{MIN\lambda}$); $L_{MAX\lambda}$ is the spectral radiance that is scaled to Q_{calmax} [W/(m^2 sr μm)] (Q_{calmax} is the maximum quantized calibrated pixel value, i.e., DN=255, corresponding to $L_{MAX\lambda}$); and DN are digital numbers of the L1 image product. Surface reflectance values (ρ) were computed by using the image-based COST method (Chavez Jr., 1996). Path radiance (L_p) values were computed by using the equation reported in Song et al. (2001) that assumes 1% surface reflectance for dark objects (Chavez Jr., 1989, 1996; Moran et al., 1992). The optical thickness for Rayleigh scattering (τ_r) was estimated according to the equation given in Kaufman (1989).

2.2. Land Cover Classification

Image classification procedures aim to automatically categorize all pixels in an image into land cover classes or themes (Lillesand et al., 2003).

Thematic mapping from remotely sensed data can be defined as grouping together cases (pixels) by their relative spectral similarity (unsupervised component) with the aim of allocating cases based on their similarity to a set of predefined classes that have been characterized spectrally (supervised component) (Foody, 2002). Multispectral images (like Landsat scenes) are frequently used to perform the classification based on spectral pattern recognition methods that exploits the pixel-by-pixel spectral information as the basis for automated classification (Lillesand et al., 2003).

Image classification involves two phases (Gong and Howarth, 1990): (1) the design of the classification scheme; and (2) the implementation of the classification scheme. Many land cover classification schemes have been developed and extensively applied in remote sensing (Anderson et al., 1976; EEA, 2006b). In this study, we mapped major land cover classes of the study inspired by the classification scheme defined by Anderson et al. (1976). Seven land cover categories representative of the study area and indicative of land change processes were considered (Table 1).

Table 1. Land cover classification scheme

ID #	Land cover class	Description
1	Bare soil	Exposed soils without vegetation or very widely spaced. Newly reclaimed areas for expansion of urban areas or roads are included.
2	Orchards	Perennial fruit orchards such as orange and lemon trees. Minor presence of herbaceous crops or pastures at the study area.
3	Shrubs	Xerophytic vegetative types with woody stems. Halophytes are highly frequent.
4	Water bodies	Lakes and irrigation reservoirs.
5	Forest land	Mediterranean coniferous evergreens dominated by *Pinus halepensis* Mill. The canopies of the trees are often scattered and generally insufficient to cover all the ground.
6	Urban/roads	Areas of intensive use, land covered by structures. Included in this category are cities, villages, transportation, power, or commercial complexes.
7	Greenhouses	Greenhouse for cultivating plants or nurseries.

The implementation of the classification schemes included the combination of supervised and unsupervised classification approaches with a spectral unmixing approach. The combination of hard classification approaches with a spectral unmixing approach has been proved highly valuable to improve the performance of land cover mapping at complex urban areas (Alberti et al., 2004). An initial unsupervised classification to map homogeneous land cover classes and to identify mixture classes was applied; pixels were then unmixed with different spectral mixture models. A final supervised classification allowed the classification of the mixture classes based on the image fraction obtained with the spectral mixture analysis. Details of the classification methods and spectral mixture analysis are provided.

2.2.1. Unsupervised Classification

An unsupervised land cover classification of Landsat images was performed with the Iterative Self-Organizing Data Analysis Technique (ISODATA). This is a well-established and frequently employed method for unsupervised classification in many remote sensing applications (Schmid et al., 2004; Shoshany, 2012).

An inherent problem to multispectral (and especially hyperspectral) remote sensing data is the high inter-band correlation, meaning that the sensor system performs in multiple measurements of the same quantity. Green et al. (1988) described a procedure called Maximum Noise Fraction transformation that provides an optimal ordering of images in terms of image quality by a series of principal component analyses to isolate noise and reduce the dimensionality of the dataset. Maximum Noise Fraction transform is commonly referred to as Minimum Noise Fraction (MNF), and it is used to detect the inherent dimensionality of image data, segregating noise from the signal in the data, and reducing computational requirements for subsequent processing tasks (Boardman and Kruse, 1994).

MNF transformation was applied to the six bands of both Landsat images. Only the first five MNF images were selected and used in subsequent analyses due to its higher quality and less noise. ISODATA parameters were iteratively adjusted to obtain up to 25 clusters. The land cover class of each cluster was assigned using aerial photographs to identify the type of land cover to which it belongs according to the classification scheme. Clusters that did not correspond exclusively to a single land cover class were identified as a mixture class. Mixture classes were masked for further analyses while pure cluster were labeled with their respective land cover class and excluded from the spectral mixture analysis.

A mixed pixel results when a sensor's Instantaneous Field of View (IFOV) includes more than one land cover type on the ground (Lillesand et al.,

2003). Spectral mixture analysis (SMA) transforms the reflectance in the bands of multispectral images to fractions of reference endmembers, which are reflectance spectra of well-characterized materials that mix to produce spectra equivalent to those of pixels of interest in the image (Adams et al., 1995). As part of SMA techniques, linear spectral unmixing (LSU) models tread the radiation recorded by a sensor as the result of a linear mixture of spectrally pure endmember radiances (Small and Lu, 2006). LSU models describe radiation reflected by an individual pixel (*i,j*) of a band *k* as the result of the product of reflectance for each land cover type by their respective mixture fraction plus an additional associated error for each pixel. The general expression of the model is presented in the following equation:

$$\rho_{i,j,k} = \sum_{m=1,p} F_{i,j,m}\, \rho_{m,k} + e_{i,j} \tag{2}$$

where ρ_{ijk} is the observed reflectance of a pixel for row *i*, column *j*, and band *k*; $F_{i,j,m}$ is the proportion of component *m* of a pixel for row *i*, column *j*, for each one of the pure components; $\rho_{m,k}$ is the characteristic reflectance for component *m* in band *k* ; and $e_{i,j}$ is the error associated to the estimation of proportions for each pixel *i, j*. The Least Square Mixing Model proposed by Shimabukuro and Smith (1991) is commonly used to resolve linear spectral mixture models. The method proposed by Shimabukuro and Smith (1991) assumes that pure endmember proportions must range between 0 and 1 and that the sum of the fractions for every component is equal to the total pixel surface.

The choice of a LSU model must consider both the landscape of the test site and the ability of the model to depict the structure, shape and distribution of the basic landscape components (Ferreira et al., 2007). Well-chosen endmembers not only represent materials found in the scene but provide an intuitive basis for understanding and describing the information in the image (Adams and Gillespie, 2006). Many authors have proposed different combinations of endmembers to analyze the spectral mixing space of urban areas. Ridd (1995) proposed a vegetation-impervious surface-soil (VIS) model for urban ecosystem analysis.

This model assumes that all pixels are a combination of different proportions of vegetation, impervious surface and soil endmembers. Small (2002) proposed a slightly different model including high albedo, low albedo and vegetation endmembers. Lu and Weng (2004) applied different combinations of three or four endmembers based on shade, green vegetation, impervious surface and dark soil endmembers. Endmembers for soil (high albedo), urban (impervious surface and high albedo), green vegetation and water (shade) covers were obtained in this study. Endmembers were image based on homogeneous areas identified by the aid of aerial ortophoto, fieldwork, and spectral indices such as the NDVI.

2.2.2. Supervised Classification

Fraction images that were derived from linear spectral unmixing can be further transformed into a hard classification or discrete classes map. Due to the fact that spectral mixture analysis was applied for each of the masked mixture clusters obtained from the ISODATA classification, different combinations of fractions images (the optimal number of fraction images was tested iteratively based on the final accuracy assessment) for each mixture cluster were obtained. A supervised classification based on a classification and regression tree (CART) originated a hard land cover map from each mixture model as was previously proposed by Lu and Weng (2004).

CART operates by recursively splitting the data until ending points, or terminal nodes, are achieved using preset criteria (Lawrence and Wright, 2001). A set of training points (30 to 50 points per class) obtained in a GIS from aerial photography visual inspection was supplied. Thus, each pixel of the masked mixture clusters was labeled with a unique land cover class. Hard land cover maps can be easily compared in change detection studies, applying map algebra operations to estimate the amount of land-cover change for the study period (1999 and 2007). The combined interpretation of fraction images of mixture components and hard land cover maps results in a powerful tool to assess land cover dynamics and human pressure on natural areas (Melendez-Pastor et al., 2010).

2.3. Accuracy Assessment and Change Detection

Pure land cover classes obtained from the ISODATA classification were combined with the hard land cover maps resulting after applying the SMA analysis and CART classification to the initial masked mixture clusters. A land cover validation database was employed to evaluate the performance of the final classification. About 600 validation points with at least 40 points for each cover class were identified. Single-pixel training led to more accurate results than other sampling methods like block training (Gong and Howarth, 1990).

Land cover map accuracy assessment was quantified with statistical methods such as the error matrix and the kappa statistic. The error matrix is a square array of numbers organized in rows and columns that express the number of sample units (i.e., pixels) assigned to a particular category relative to the actual category as indicated by the reference data (Congalton, 2004). Reference data are in the columns, while the rows indicate the map categories to be assessed. This form of expressing accuracy as an error matrix is an effective way to evaluate both errors of inclusion (commission errors) and errors of exclusion (omission errors) present in the classification as well as the overall accuracy (Congalton et al., 1983). In addition to the error matrix, the Kappa coefficient developed by Cohen (1960) was employed to quantify the accuracy of the land cover map. Cohen's Kappa (or KHAT) is a measure of agreement for nominal scales based on the difference between the actual agreement of the classification (i.e., agreement between computer classification and reference data as indicated by the diagonal elements) and the chance agreement, which is indicated by the product of the row and column marginal (Congalton et al., 1983).

3. Results and Discussion

3.1. Mixed Classes Delineation and Spectral Mixture Analysis

The use of the unsupervised classification computed 25 and 24 clusters for 1999 and 2007, respectively. Using aerial photographs and fieldwork, land cover classes of the clusters were assigned. The clusters that could not be allocated uniquely to a land cover class were identified as mixture classes (Figure 2.b).

For the image of 1999, in the study area, 71.3% was uniquely assigned to a single land cover class with the ISODATA unsupervised classification. However, three mixture clusters were identified according to the following mixed classes:

1. Bare soil and urban/roads: 25.23% of the study area.
2. Bare soil, water and greenhouses: 1.04% of the study area.
3. Coniferous forest, shrubs and water: 2.42% of the study area.

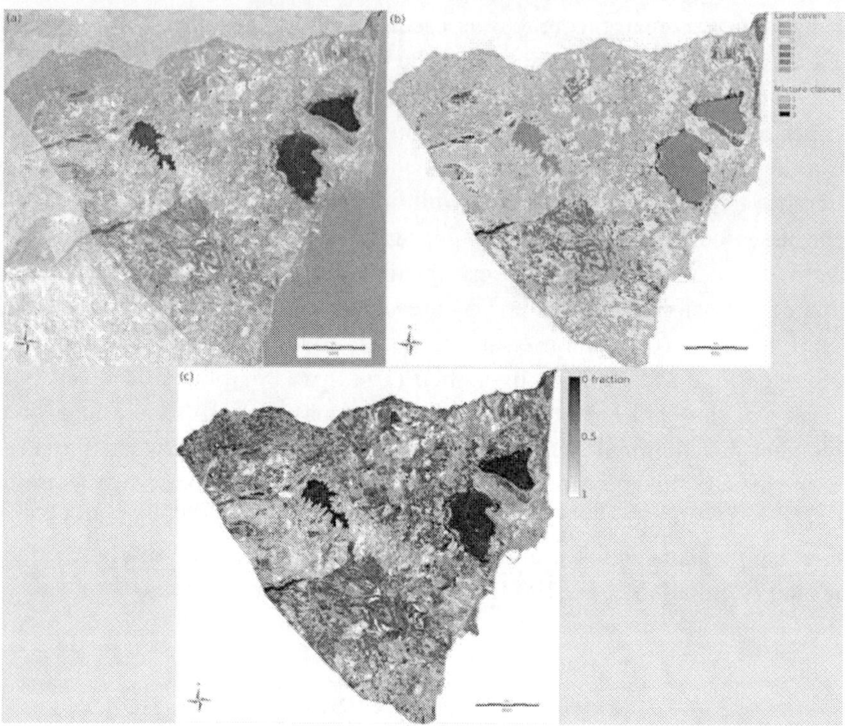

Figure 2. Example of digital image processing stages for 1999: (1) initial false color composite RGB:742; (2) ISODATA unsupervised classification result, note the three mixture classes; and (3) soil endmember fraction image.

In 2007, of the study are, 95.5% was uniquely assigned to a single land cover class with the ISODATA unsupervised classification. Only two mixture clusters were identified according to the following mixed classes:

1. Bare soil and urban/roads: 3.80% of the study area.
2. Bare soil, water and greenhouses: 0.71% of the study area.

A mask for each mixture class was defined and employed in the spectral mixture analysis, while pure clusters were excluded up to the final accuracy assessment.

Four endmembers were defined for 1999 and 2007, according to the following criteria (Figure 3): (1) soil: bare soil areas at agricultural fields or new land reclamation areas for future building (Figure 2.c); (2) urban: residential areas and communication infrastructures; (3) vegetation: dense vegetation areas such as orchards or forest land with NDVI > 0.7; and (4) water: large water bodies such as the Pedrera reservoir and the Lagunas de la Mata and Torrevieja wetlands. Spectral mixture analysis was applied to different combinations of endmembers, from two to four endmembers. Michishita et al. (2012) also tested a huge number of endmember combinations of green vegetation, non-photosynthetic vegetation and soil, built-up areas and different water classes in the Poyang Lake area (China).

They tested from two endmember models to four endmember models. Optimal endmembers choice was done after applying the supervised classification and the accuracy assessment as an iterative process. Therefore, three spectral mixture models selected for each of the ISODATA mixture clusters were: (a) mixture 1: a spectral mixture model with soil and urban endmembers to discriminate between urban/roads and bare soil land cover classes; (b) mixture 2: a spectral mixture model with soil, urban and water endmembers to discriminate among bare soil, water and greenhouses land cover classes; and (c) mixture 3: a spectral mixture model with soil, vegetation and water endmembers to discriminate among coniferous forest, shrubs and water land cover classes.

In relation to the selection of endmembers for the discrimination of mixed classes, Alberti et al. (2004) employed a paved, vegetation a shade mixture model for mixed urban areas, and vegetation, bare soil and shade mixture model for forest-grass discrimination at Seattle metropolitan area (USA).

Lu and Weng (2004, 2006) proposed a combination of shade, green vegetation and soil/impervious surface to characterize the urban landscape of Indianapolis (USA).

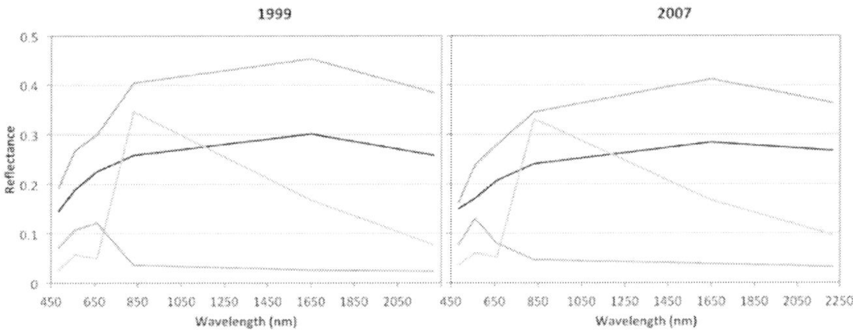

Figure 3. Soil (red), urban/roads (black), vegetation (green) and water (blue) endmembers for 1999 and 2007, spectral mixture analysis.

3.2. Land Cover Classification and Accuracy Assessment

A supervised classification was applied to the endmember fraction images at the masked areas. A different regression tree was developed for each mixture model. A set of training points (30 to 50 points per class) obtained in a GIS from aerial photography visual inspection was employed training the CART algorithm.

Accuracy assessment was developed with a different dataset in order to discriminate the most successful spectral mixture models and the optimal parameters of the CART algorithm. A 3 x 3 validation pixels window size was employed (Alberti et al., 2004). After several iterations, final land cover maps were obtained by merging the pure classes of the ISODATA unsupervised classification with the land cover maps obtained for the mixture clusters masked areas (Figure 4). Clump and sieve procedures with a 3 x 3 moving window were applied in order to eliminate spurious pixels.

The accuracy assessment of the final land cover maps was satisfactory enough, with an overall accuracy of 86.51% for the 1999 (Table 2), and 89.38% for the 2007 (Table 3). A total of 608 validation points were employed for the 1999 land cover map, while 584 validation points were employed for the 2007 land cover map. The Kappa coefficient values were 0.8362 for 1999, and 0.8693 for 2007.

Table 2. Accuracy assessment for 1999 land cover classification

		Observed in 1999								
		Class 1	Class 2	Class 3	Class 4	Class 5	Class 6	Class 7	TOTAL	User's Accuracy
Classified in 1999	Class 1	100	1	5	0	0	9	3	118	85%
	Class 2	3	94	1	0	0	3	2	103	91%
	Class 3	3	1	34	0	1	29	0	68	50%
	Class 4	0	0	0	63	0	0	0	63	100%
	Class 5	0	0	11	0	50	1	0	62	81%
	Class 6	5	0	0	0	0	150	0	155	97%
	Class 7	1	0	2	0	0	1	35	39	90%
	TOTAL	112	96	53	63	51	193	40	608	
	Producer's Accuracy	89%	98%	64%	100%	98%	78%	88%		
Overall accuracy = 86.51%		Kappa coefficient= 0.8362								

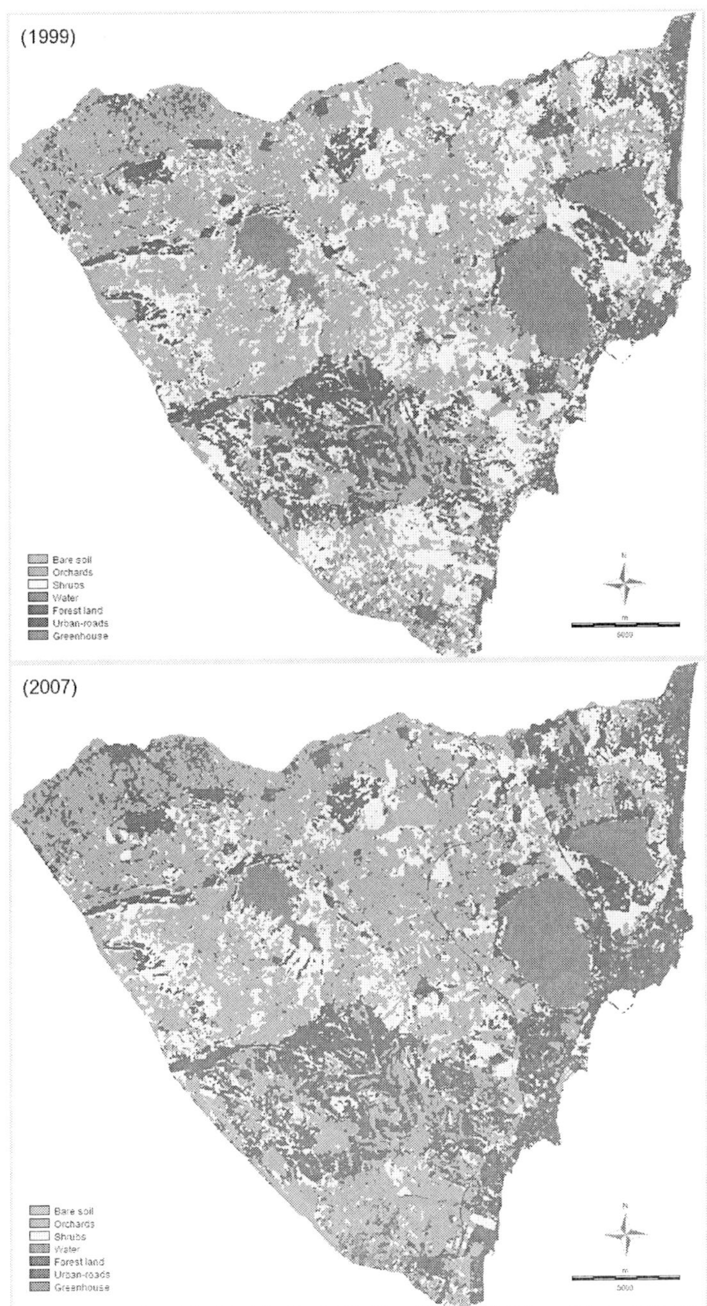

Figure 4. Land cover classification maps for 1999 and 2007.

Table 3. Accuracy assessment for 2007 land cover classification

	Observer in 2007								
Classified in 2007	Class 1	Class 2	Class 3	Class 4	Class 5	Class 6	Class 7	TOTALS	User's Accuracy
Class 1	79	6	4	0	0	6	9	104	76%
Class 2	0	90	0	0	0	0	0	90	100%
Class 3	1	0	33	0	0	10	0	44	75%
Class 4	0	0	0	62	0	0	0	62	100%
Class 5	0	0	16	0	51	6	0	73	70%
Class 6	1	0	0	1	0	176	0	178	99%
Class 7	0	0	0	0	0	2	31	33	94%
TOTALS	81	96	53	63	51	200	40	584	
Producer's Accuracy	98%	94%	62%	98%	100%	88%	78%		
Overall accuracy = 89.38%		Kappa coefficient = 0.8693							

Table 4. Land cover classes surface (ha) for 1999, 2007, and temporal evolution

Land cover classes	1999 (ha)	2007 (ha)	Change 2007-1999 (ha)
Bare soil	8627.04	10411.38	1784.34
Orchards	15465.60	13635.09	-1830.51
Shrubs	14287.05	10268.01	-4019.04
Water	3463.11	3177.63	-285.48
Coniferous forest	8100.45	8085.33	-15.12
Urban roads	3509.73	7970.94	4461.21
Greenhouse	378.81	283.41	-95.40

According to Landis and Koch (1977), KHAT values above 0.8 exhibit a good agreement, and thus our maps are confident enough. The lowest producer and user accuracy values were reported for shrubs (class 3) that were confused with coniferous forest (class 5) or even urban areas (class 6).

Similar overall agreement (0.8-0.9) and Kappa coefficient (0.7-0.9) values were obtained with a similar approaches combining hard classification and spectral mixture analysis as developed by Alberti et al. (2004) and Lu and Weng (2004). Thus, land cover maps derived from the spectral mixture analysis have reasonably high accuracy and are sufficient for urban landscape analysis and growth detection (Lu and Weng, 2004).

Dominant land cover classes in the study area were orchards, shrubs and bare soil, which accounted a 71.3% of the study area for 1999. The contribution of those land cover classes decreased to a 63.7% of the study area for 2007. A huge increase of urban/road areas (+127%) was in detriment of shrubs areas. The decrease of orchard areas was partially explained by agricultural land consumption for urban roads and for the replanting processes. Bare soil areas increased as consequence of new urban-planned areas. No significant surface variations were denoted for the other land cover classes.

Urban growth is a complex phenomenon that promotes negative impacts on soil resources. Navarro-Pedreño et al. (2012) evaluated the impact of urban growth in soil resource in the nearby municipality of Elche. They observed that urban growth in coastal areas was the dominant urban growth process in the last decades. Major urban expansion in our study area was focused around the city of Torrevieja and along the coastline. New residential areas extensively occupied the borders of the Lagunas de la Mata and Torrevieja Natural Park.

Moreover, new residential areas were also developed in the north of the study area, which correspond with alluvial plains traditionally devoted to the irrigation agriculture. Navarro-Pedreño et al. (2012) also noted the severe impact of coastal urbanization at moderate to high land capability soils for agricultural activities.

CONCLUSION

The methodological approach combining unsupervised (ISODATA) and supervised (CART) classification with spectral mixture analysis to obtain land cover maps was successful, and land change was detected in this complex area. The accuracy assessment of the final land cover maps reported an overall accuracy of 86.51% for 1999, and 89.38% for 2007, and the detection of soil sealing increment was adequately determined. The Kappa coefficient values were 0.8362 for 1999, and 0.8693 for 2007, thus providing a good confidence in the land cover maps for a proper multi-temporal analysis and showing the goodness of this methodology.

In this study, major land cover change was detected in the increase of urban areas (+ 127%) and the diminution of shrubs and orchard areas (both natural and agricultural spaces). A large demand for residential houses beside the sea and the importance of the coastal tourism promoted the urban growth. Huge soil sealing processes are expected with very negative impact on the soil resource due to this irreversible process. Optimal land use planning is needed to minimize the impact of the massive development of new residential areas along the coastline and around natural areas like the RAMSAR wetland of the Lagunas de la Mata y Torrevieja Natural Park, and this methodology may be an adequate tool to determine the historical changes and the major areas affected by soil sealing and could help in future land planning strategies.

Acknowledgments

The authors would like to thank the U.S. Geological Survey for providing the Landsat satellite imagery used in the study.

References

Adams, J. B., Gillespie, A. R. (2006). *Remote Sensing of Landscapes with Spectral Images. A Physical Modeling Approach.* Cambridge, UK: Cambridge University Press.

Adams, J. B., Sabol, D. E., Kapos, V., Filho, R. A., Roberts, D. A., Smith, M. O., Gillespie, A. R. (1995). Classification of Multispectral Images Based on Fractions of Endmembers: Application to Land-Cover Change in the Brazilian Amazon. *Remote Sensing of Environment*, 52: 137-154.

Alberti, M., Weeks, R., Coe, S. (2004). Urban Land-Cover Change Analysis in Central Puget Sound. *Photogrammetric Engineering and Remote Sensing*, 70(9): 1043-1052.

Anderson, J. R., Hardy, E. E., Roach, J. T., Witmer, R. E. (1976). A Land Use and Land Cover Classification System For Use With Remote Sensor Data. Geological Survey Professional Paper 964. Washington D.C., USA.

Boardman, J. W., Kruse, F. A. (1994). Automated spectral analysis: A geological example using AVIRIS data, northern Grapevine Mountains, Nevada. Proceedings, ERIM Tenth Thematic Conference on Geologic Remote Sensing (p. I-407 - I-418). Ann Arbor (MI), USA: Environmental Research Institute of Michigan.

Chander, G., Markham, B. (2003). Revised Landsat-5 TM Radiometric Calibration Procedures and Postcalibration Dynamic Ranges. *IEEE Transactions on Geoscience and Remote Sensing*, 41(11): 2674-2677.

Chander, G., Haque, M. O., Micijevic, E., Barsi, J. A. (2010). A Procedure for Radiometric Recalibration of Landsat 5 TM Reflective-Band Data. *IEEE Transactions on Geoscience and Remote Sensing,* 48(1): 556-574.

Chavez Jr, P. S. (1989). Radiometric calibration of Landsat Thematic Mapper multispectral images. *Photogrammetric Engineering and Remote Sensing*, 55(9): 1285 -1294.

Chavez Jr, P. S. (1996). Image-based atmospheric corrections—Revisited and improved. *Photogrammetric Engineering and Remote Sensing*, 62(9):1025-1036.

Cohen, J. (1960). A Coefficient of Agreement for Nominal Scales. *Educational and Psychological Measurement*, 20(1): 37-46.

Congalton, R. G. (2004). Putting the map back in map accuracy assessment. In R. S. Lunetta, J. G. Lyon (Eds.), *Remote Sensing and GIS accuracy assessment* (pp. 1-11). Boca Raton (FL), USA: CRC Press.

Congalton, R. G., Oderwald, R. G., Mean, R. A. (1983). Assessing Landsat Classification Accuracy Using Discrete Multivariate Analysis Statistical Techniques. *Photogrammetric Engineering and Remote Sensing*, 49(12): 1671-1678.

EEA. (2006a). Urban sprawl in Europe. The ignored challenge. EEA Report 10/2006. Luxembourg.

EEA. (2006b). Corine Land Cover classes. European Topic Centre on Land Use and Spatial Information (EIONET). Retrieved March 14, 2011, URL: http://sia.eionet.europa.eu/CLC2000/classes/index_html.

European Commission. (2002). Towards a Thematic Strategy for Soil Protection. COM(2002) 179 final. Brussels, Belgium: Commission of the European Communities.

Ferreira, M. E., Ferreira, L. G., Sano, E. E., Shimabukuro, Y. E. (2007). Spectral linear mixture modelling approaches for land cover mapping of tropical savanna areas in Brazil. *International Journal of Remote Sensing*, 28(2): 413-429.

Foody, G. M. (2002). Status of land cover classification accuracy assessment. *Remote Sensing of Environment*, 80(1): 185-201.

Gong, P., Howarth, P. (1990). An Assessment of Some Factors Influencing Multispectral Land-Cover Classification. *Photogrammetric Engineering and Remote Sensing,* 56(5): 597-603.

Green, A. A., Berman, M., Switzer, P., Craig, M. D. (1988). A transformation for ordering multispectral data in terms of image quality with implications for noise removal. *IEEE Transactions on Geoscience and Remote Sensing*, 26(1): 65-74.

Irish, R. R. (2012). Landsat 7 Science Data Users Handbook. Landsat Project Science Office, Goddard Space Flight Center-NASA. Retrieved July 17, 2012. URL: http://landsathandbook.gsfc.nasa.gov.

Jensen, J. R. (2005). *Introductory Digital Image Processing (3rd Edition).* Upper Saddle River (NJ), USA: Prentice Hall.

Jensen, J. R. (2007). *Remote Sensing of the Environment: An Earth Resource Perspective*. Upper Saddle River (NJ), USA: Prentice Hall.

Kaufman, Y. J. (1989). The atmospheric effect on remote sensing and its corrections. In G. Asrar (Ed.), *Theory and Applications of Optical Remote Sensing* (pp. 336-428). New York, USA: Wiley-Interscience.

Landis, J. R., Koch, G. G. (1977). The measurement of observer agreement for categorical data. *Biometrics,* 33: 159-174.

Lawrence, R. L., Wright, A. (2001). Rule-Based Classification Systems Using Classification and Regression Tree (CART) Analysis. *Photogrammetric Engineering and Remote Sensing,* 67(10): 1137-1142.

Lillesand, T. M., Kiefer, R. W., and Chipman, J. W. (2003). *Remote Sensing and Image Interpretation (5th ed.)*. Hoboken (NJ), USA: John Wiley and Sons.

Lu, D, Weng, Q. (2004). Spectral Mixture Analysis of the Urban Landscape in Indianapolis with Landsat ETM+ Imagery. *Photogrammetric Engineering and Remote Sensing*, 70(9): 1053-1062.

Mather, P. M. (2004). *Computer Processing of Remotely-Sensed Images: An Introduction (3rd ed.)*. Chichester, UK: Wiley.

Melendez-Pastor, I., Navarro-Pedreño, J., Gómez, I., Koch, M. (2010). Detecting drought induced environmental changes in a Mediterranean wetland by remote sensing. *Applied Geography*, 30(2): 254-262.

Mermut, A. R., and Eswaran, H. (2001). Some major developments in soil science since the mid-1960s. *Geoderma*, 100(3-4): 403-426.

Michishita, R., Jiang, Z., Xu, B. (2012). Monitoring two decades of urbanization in the Poyang Lake area, China through spectral unmixing. *Remote Sensing of Environment*, 117: 3-18.

Moran, M. S., Jackson, R. D., Slater, P. N., Teillet, P. M. (1992). Evaluation of simplified procedures for retrieval of land surface reflectance factors from satellite sensor output. *Remote Sensing of Environment*, 41: 169-184.

Navarro Pedreño, J., Meléndez-Pastor, I., Gómez Lucas, I. (2012). Impact of three decades of urban growth on soil resources in Elche. *Spanish Journal of Soil Science*, 2(1): 55-69.

Nizeyimana, E. (2006). Remote sensing and GIS integration. In R. Lal (Ed.), *Encyclopedia of Soil Science, Second Edition 2 Volume Set* (pp. 1469-1472). New York, USA: Taylor and Francis Group.

Prokop, G., Jobstmann, H., Schönbauer, A. (2011). *Overview of best practices for limiting soil sealing or mitigating its effects in EU-27.* Wien, Austria: Environment Agency Austria.

Ridd, M. K. (1995). Exploring a V-I-S (vegetation-impervious surface-soil) model for urban ecosystem analysis through remote sensing: comparative anatomy for cities. *International Journal of Remote Sensing*, 16(12): 2165-2185.

Schmid, T., Koch, M., Gumuzzio, J., Mather, P. M. (2004). A spectral library for a semi-arid wetland and its application to studies of wetland degradation using hyperspectral and multispectral data. *International Journal of Remote Sensing*, 25(13): 2485-2496.

Shimabukuro, Y. E., Smith, J. A. (1991). The least-squares mixing models to generate fraction images derived from remote sensing multispectral data. *IEEE Transactions on Geoscience and Remote Sensing*, 29(1): 16-20.

Shoshany, M. (2012). Identifying Desert Thresholds by Mapping Inverse Erodibility and Recovery Potentials in Patch Patterns Using Spectral and Morphological Algorithms. *Land Degradation and Development*, doi:10.1002/ldr.2146.

Small, C. (2002). Multitemporal analysis of urban reflectance. *Remote Sensing of Environment*, 81: 427-442.

Small, C., Lu, J. W. T. (2006). Estimation and vicarious validation of urban vegetation abundance by spectral mixture analysis. *Remote Sensing of Environment,* 100(4): 441-456.

Song, C., Woodcock, C. E., Seto, K. C., Lenney, M. P., Macomber, S. A. (2001). Classification and Change Detection Using Landsat TM Data: When and How to Correct Atmospheric Effects? *Remote Sensing of Environment,* 75(2): 230-244.

Weng, Qihao, Lu, D., Liang, B. (2006). Urban Surface Biophysical Descriptors and Land Surface Temperature Variations. *Photogrammetric Engineering and Remote Sensing*, 72(11): 1275-1286.

INDEX

A

B

C

D

E

F

G

H

I

J

K

L

M

N

O

P

Q

R

S

T

U

V

W

Y

Z